Climate Change and Questions of Justice

ISBN 1-60123-187-3/ 978-1-60123-187-1

Acknowledgments

Climate Change and Questions of Justice was developed by the Choices Program with the assistance of faculty at the Watson Institute for International and Public Affairs, scholars at Brown University, and other experts in the field. We wish to thank the following researchers for their invaluable input:

DAVID CIPLET
Assistant Professor of Environmental Studies
The University of Colorado at Boulder

GUY EDWARDS
Research Fellow, Codirector of Climate Development Lab
Institute at Brown for Environment and Society
Brown University

TIMOTHY D. HERBERT
Henry L. Doherty Professor of Oceanography
Professor and Chair of Earth, Environmental, and Planetary
Sciences Department
Institute at Brown for Environment and Society
Brown University

MICHAEL WARREN MURPHY
National Science Foundation IGERT Fellow, PhD Candidate
Watson Institute for International and Public Affairs
Brown University

J. TIMMONS ROBERTS
Ittleson Professor of Environmental Studies and Sociology
Institute at Brown for Environment and Society
Brown University

DOV SAX
Associate Professor of Ecology and Evolutionary Biology
Associate Professor of Environment and Society
Deputy Director (Teaching), Institute at Brown for
Environment and Society, Brown University

LEAH K. VANWEY
Professor of Environment and Society and Sociology
Associate Provost for Academic Space
Brown University

Thank you to the Climate Development Lab of Brown University (Jeffrey Baum, Cassidy Bennett, Camila Bustos, Ximena Carranza, Alexis Durand, Victoria Hoffmeister, Zihao Jiang, Maris Jones, Alison Kirsch, Sophie Purdom, Allison Reilly, and Marguerite Suozzo-Gole) for their guidance and feedback on this curriculum unit.

Thank you to Leah Elliott, Jessica Fields, Danielle Johnstone, and Maya Lindberg for their help in developing and writing this unit.

Maps by Alexander Sayer Gard-Murray.

Cover photograph by the NOAA Photo Library (CC BY 2.0).

The Choices Program

Director
Susan Graseck

Curriculum Development Director
Andy Blackadar

Professional Development Director
Mimi Stephens

Curriculum Development Assistant Director
Susannah Bechtel

Curriculum Writer
Lindsay Turchan

Research and Writing Intern
Ada Okun

Administrative Manager
Kathleen Magiera

Marketing and Social Media Manager
Jillian McGuire Turbitt

Manager, Digital Media Group
Tanya Waldburger

Office Assistant
Lisa Blake

Contents

Cumulative Global Carbon Emissions

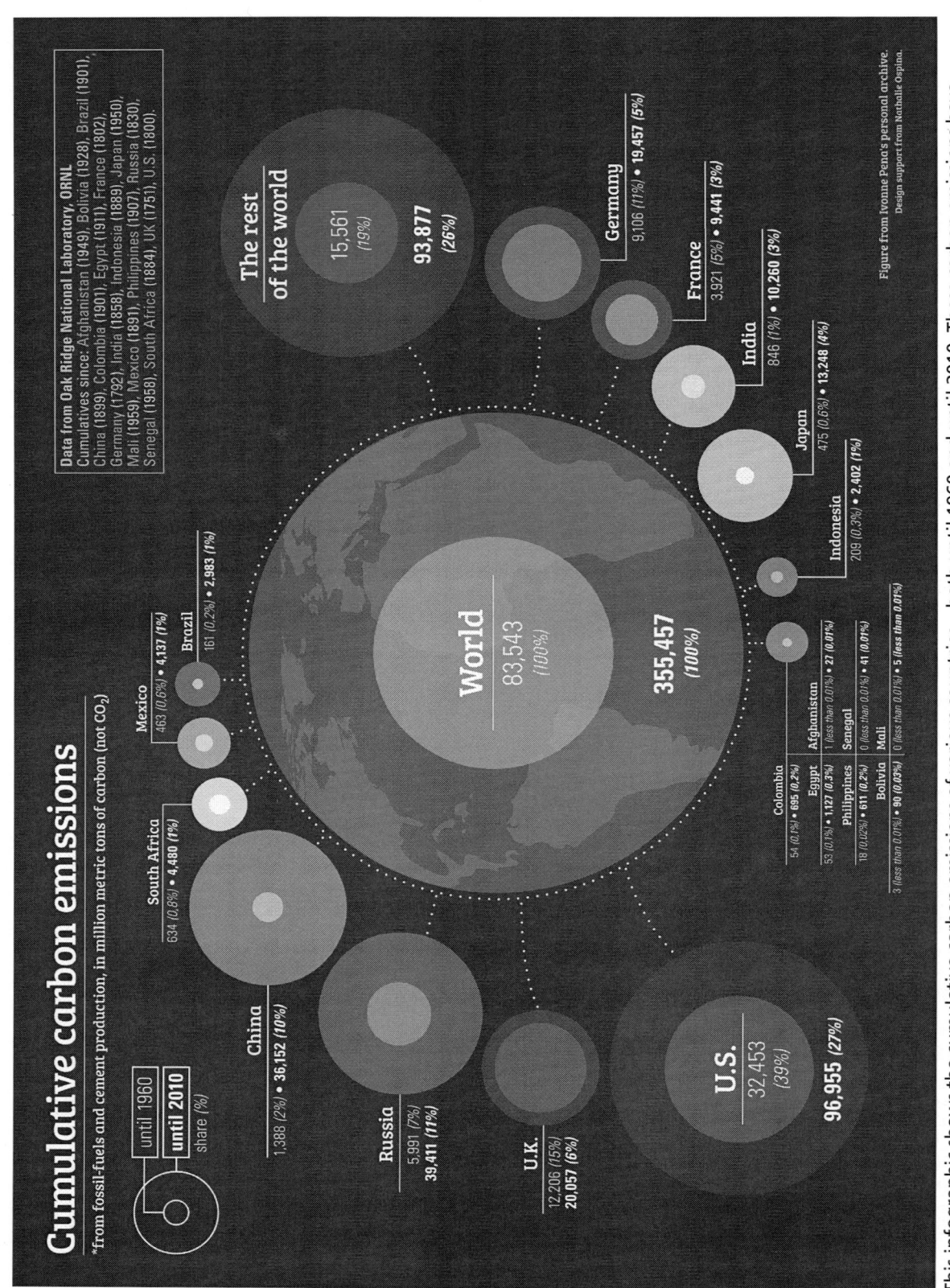

Cumulative carbon emissions
*from fossil-fuels and cement production, in million metric tons of carbon (not CO₂)

until 1960
until 2010
share (%)

Data from Oak Ridge National Laboratory, ORNL
Cumulatives since: Afghanistan (1949), Bolivia (1928), Brazil (1901), China (1899), Colombia (1901), Egypt (1911), France (1802), Germany (1792), India (1858), Indonesia (1889), Japan (1950), Mali (1959), Mexico (1891), Philippines (1907), Russia (1830), Senegal (1958), South Africa (1884), UK (1751), U.S. (1800).

The rest of the world
15,561 (19%)
93,877 (26%)

Germany
9,106 (11%) • **19,457 (5%)**

France
3,921 (5%) • **9,441 (3%)**

India
846 (1%) • **10,260 (3%)**

Japan
475 (0.6%) • **13,248 (4%)**

Indonesia
209 (0.3%) • **2,402 (1%)**

Mexico
463 (0.6%) • **4,137 (1%)**

Brazil
161 (0.2%) • **2,983 (1%)**

South Africa
634 (0.8%) • **4,480 (1%)**

World
83,543 (100%)
355,457 (100%)

China
1,388 (2%) • **36,152 (10%)**

Russia
5,991 (7%) • **39,411 (11%)**

U.K.
12,206 (15%) • **20,057 (6%)**

U.S.
32,453 (39%)
96,955 (27%)

Colombia
54 (0.1%) • **695 (0.2%)**

Egypt
53 (0.1%) • **1127 (0.3%)**

Afghanistan
1 (less than 0.01%) • **27 (0.01%)**

Philippines
18 (0.02%) • **611 (0.2%)**

Senegal
0 (less than 0.01%) • **41 (0.01%)**

Bolivia
3 (less than 0.01%) • **90 (0.03%)**

Mali
0 (less than 0.01%) • **5 (less than 0.01%)**

Figure from Ivonne Peña's personal archive.
Design support from Nathalie Ospina.

This infographic shows the cumulative carbon emissions of various countries—both until 1960 and until 2010. These carbon emissions have contributed to global climate change.

Introduction: The Challenge of a Unified Response

"Coming here today, I have no hidden agenda. I am fighting for my future. Losing my future is not like losing an election or a few points on the stock market. I am here to speak for all generations to come. I am here to speak on behalf of the starving children around the world whose cries go unheard. I am here to speak for the countless animals dying across this planet because they have nowhere left to go. We cannot afford to be not heard."

—Severn Suzuki, speaking on behalf of the Environmental Children's Organization at the UN's Earth Summit, 1992

After addressing delegates of governments from across the world, Severn Suzuki became known as "the girl who silenced the world in five minutes." Thirteen-year-old Suzuki and three of her peers had raised money to attend the Earth Summit in Rio de Janeiro, Brazil, where leaders and officials of 172 countries were meeting to establish an agenda to address global environmental issues. At the Summit, Suzuki stood before this vast audience of international power-holders and urged them to consider the futures of their children—the futures of young people like her.

Ultimately, the Earth Summit resulted in 165 governments agreeing that climate change was a shared and dangerous problem. As of 2017, 196 countries have signed the United Nations Framework Convention on Climate Change (UNFCCC), promising to work together to reduce or prevent increases in the amounts of greenhouse gases (the gases that cause climate change) in the atmosphere. The UNFCCC set in motion a series of climate change conferences that continue to this day.

Developing responses to climate change that are acceptable to all members of the international community is no easy task. The challenge is to take into account the many different concerns of countries, ordinary people, businesses, and activists in creating an effective set of policies to address this shared problem. While scientists argue that human-caused climate change is an urgent matter, policy makers disagree about the severity of the threat and how to respond.

These disagreements have made it difficult for national governments to develop a unified response to climate change, even after more than twenty years of meetings and conferences. The slow pace of this process has led people and groups outside of national governments to develop their own responses. Just as Severn Suzuki, with her 1992 speech, challenged the idea that only national leaders have a stake in environmental issues, organizations and individuals are finding ways to raise their voices and create change. Local governments around the world are designing plans to help

Country delegates at a 2014 United Nations (UN) Climate Change Conference in Bonn, Germany.

Jan Golinski, UNclimatechange (CC BY 2.0).

People marching outside a 2009 UN Climate Change Conference in Copenhagen, Denmark.

their communities adapt to the new conditions caused by climate change. Several state governments have voluntarily adopted stricter environmental standards. Nongovernmental organizations (NGOs) work to influence policy and educate the public. Corporations are seeing business opportunities in providing more environmentally friendly products.

In the coming pages, you will explore the pressing need for an effective response to climate change and take on the challenge of determining what that response should be. You will begin by examining the causes and effects of climate change and by analyzing the efforts to respond to this global problem. You will explore eight case studies that show how different parts of the world are experiencing a changing climate. The readings highlight many of the issues that make developing a unified

response to climate change so difficult. As you read, keep these questions in mind:

- How does climate change affect different regions of the world?

- Who is vulnerable to climate change?

- Who is responsible for climate change?

- How can the international community respond to climate change in a fair and effective way?

After completing your readings, you will have a chance to grapple with these same questions during a climate conference simulation with your classmates. You will take on the roles of national leaders, representatives of NGOs, and technical experts to debate and discuss questions of climate justice.

Part I: The Causes and Effects of Global Climate Change

Understanding Climate Change

Wherever people live, they become familiar with their local climate. Some may live in regions where it snows or rains frequently, places where summers are hot and dry, or in tropical or arid regions. Weather, rainfall, temperature, and human activity affect the types of plants, trees, and animals that live in a region and contribute to the local climate.

What is climate change?

The term "climate change" can refer to any significant shifts in temperature, rainfall, wind, and other environmental factors that occur over decades or more. The earth's climate has undergone natural variations throughout the entire history of the planet. Today, the climate change we hear about most often refers to changes caused primarily by human activity that alters the composition of the atmosphere.

The earth's atmosphere is made up of numerous gases that make life possible. Gases such as carbon dioxide (CO_2), methane, nitrous oxide, and water vapor exist naturally in the atmosphere and warm the earth to a temperature at which humans can live. These gases make up only a small percentage of the atmosphere. They are called "greenhouse gases" because they trap heat in the atmosphere by absorbing energy that would otherwise be radiated back into space. The process works the same way that a greenhouse for plants prevents heat from escaping beyond its glass panels.

To get a sense of how important these gases are to life on Earth, we can look

> ### Part I Definitions
>
> **Climate change**—Any significant shifts in temperature, rainfall, wind, and other environmental factors that occur over decades or more.
>
> **Greenhouse gases**—Gases in the atmosphere that warm the earth.
>
> **Climate change refugees**—People who are forced to flee their homes due to the effects of climate change.

at how they affect temperature. The earth's current average temperature is 59°F. Without greenhouse gases, the earth's average temperature would drop to around 0°F, potentially making the planet cold enough for all water on Earth to freeze.

Mattes (public domain via Wikimedia Commons).

The roof and glass walls of a greenhouse let the sun's energy in and keep heat trapped inside, creating an environment warm enough for certain plants to grow. Gases like CO_2 and methane in the Earth's atmosphere are called greenhouse gases. This is because they perform a similar function to the glass of a greenhouse: allowing in sunlight and trapping heat.

Over the past 150 years, human activity—primarily the burning of fossil fuels—has increased the amount of greenhouse gases in the atmosphere. With more greenhouse gases, the earth gets warmer, which is why climate change today is often referred to as "global warming."

Often, the topic of climate change will come up after an extreme weather event like a hurricane or a blizzard. When there is a long heat wave or a series of powerful storms, people might think that they are witnessing climate change. But while it is tempting to attribute these weather patterns to global warming, in truth, no one can say climate change is happening based on their own observations over a few days, months, or even years. Scientific data collected over a period of decades has led to the conclusion that the earth's climate is drastically changing.

> **"Even with climate change, you will occasionally see cooler-than-normal summers or a typically cold winter. Don't let that fool you."**
>
> —James E. Hansen, director of the NASA Goddard Institute for Space Studies, 2012

What is causing climate change?

Humans' use of fossil fuels (coal, oil, and natural gas) produces CO_2 and is the leading cause of climate change. Fossil fuels were formed from plants and animals that lived millions of years ago.

In ages past, their remains were buried deep within the earth's crust and were transformed into petroleum and natural gas by intense heat and pressure. Since the start of the Industrial Revolution in the late eighteenth century, human consumption of fossil fuels has soared, and the amount of CO_2 in the atmosphere has increased by more than 40 percent.

Some scientists say that humans' impact on the climate traces back to the origins of agriculture. As farming replaced hunting and gathering as the dominant means of survival for humans, people cleared land of trees so that it could be used to grow crops or raise

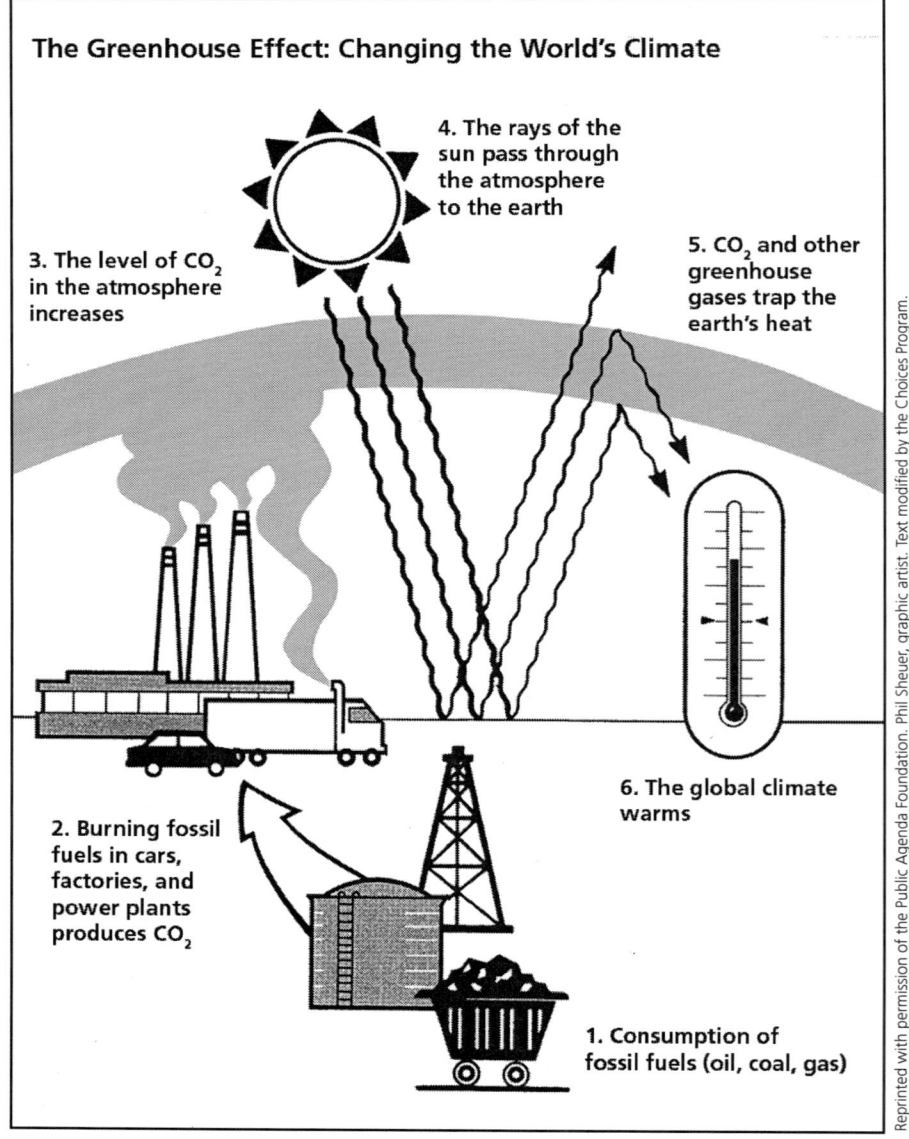

The Greenhouse Effect: Changing the World's Climate

3. The level of CO_2 in the atmosphere increases

4. The rays of the sun pass through the atmosphere to the earth

5. CO_2 and other greenhouse gases trap the earth's heat

6. The global climate warms

2. Burning fossil fuels in cars, factories, and power plants produces CO_2

1. Consumption of fossil fuels (oil, coal, gas)

Reprinted with permission of the Public Agenda Foundation. Phil Sheuer, graphic artist. Text modified by the Choices Program.

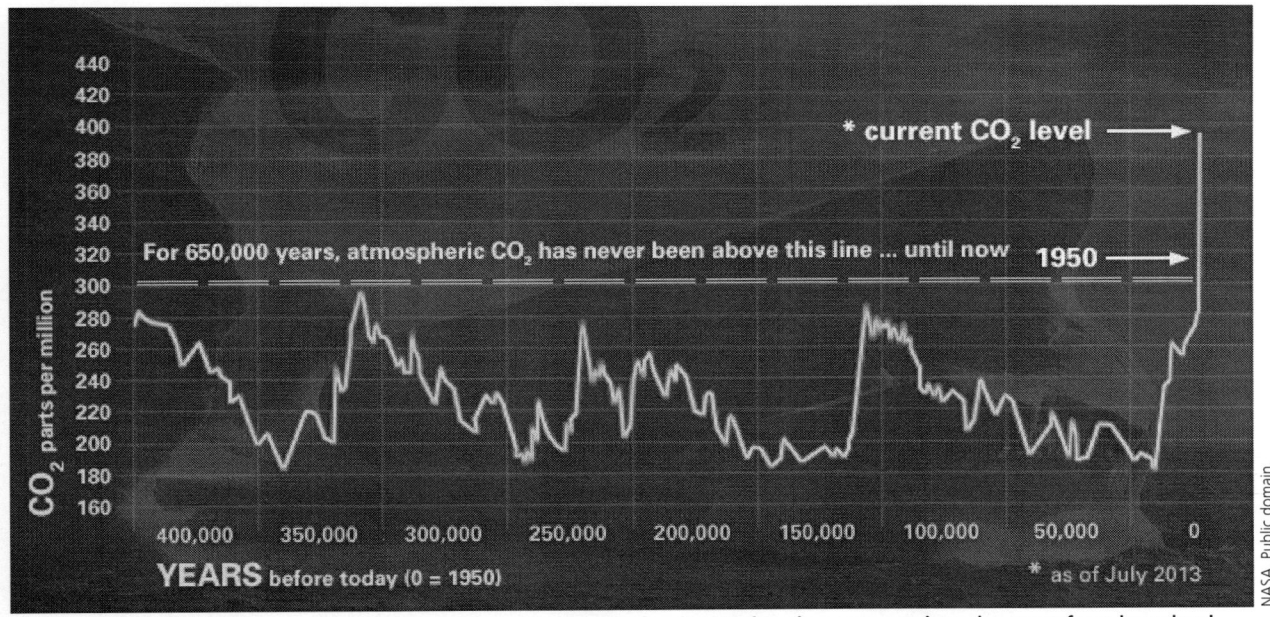

NASA. Public domain.

This graph shows the recent dramatic increase in atmospheric CO_2 levels compared to the past four hundred thousand years.

livestock. Farmland now takes up almost half of the earth's land surface, replacing what were once vast expanses of forest and woodlands. Because trees absorb CO_2, the deforestation meant that less CO_2 was being removed from the atmosphere. As a result, the percentage of greenhouse gases in the atmosphere gradually increased. However, it was not until humans began emitting large amounts of greenhouse gases by burning fossil fuels for transportation, industry, heat, and electricity that this increase became pronounced.

How did industrialization contribute to the use of fossil fuels?

With the beginning of industrialization in the mid 1700s, Britain was the first country to replace wood with fossil fuels as its main source of energy. In addition to clearing forests, people used increasing amounts of coal to meet Britain's ever-growing appetite for energy. In the 1800s, industrialists expanded coal mining and developed oil-drilling techniques. By the turn of the twentieth century, the United States had taken the lead in forging an industrial economy powered by coal and oil. Today, these two fuel sources, along with natural gas, supply roughly 80 percent of the world's energy. In 2012, coal alone was responsible for 43 percent of the total CO_2 emissions from human activity.

Courtesy of Leonard Bentley (CC BY-SA 2.0 via Flickr).

This photograph shows the British Houses of Parliament in the early twentieth century. Visible at the left of the image, factories belching smoke into the atmosphere were only yards away.

Carbon dioxide accounts for the majority of the greenhouse gas emissions from human activity. On the one hand, CO_2 is essential for life on Earth. For example, plants require it just as animals need oxygen. But it is the large and rapid increase in CO_2 emissions from human activity that is causing the earth to warm more now than it has in the past.

What is the evidence of climate change?

Research shows that the earth's average surface temperature has risen 1.4°F since 1880. This may not seem like a large change, but it is significant considering it has happened about ten times faster than the average rate of change after each ice age.

Most of this warming has taken place since 1970. The rapid rate of change concerns scientists. The overwhelming agreement among scientists that the climate is drastically changing is based on rising temperatures and other evidence as well. Ocean temperatures and acidity levels have risen, glaciers (large masses of slowly moving ice) around the world are shrinking in size, and extreme weather events have increased in frequency.

How is climate science produced?

Science is sometimes thought to be a set of permanent facts. But the body of scientific knowledge is ever-changing as scientists continually work to refine understandings of the natural world. This is also true of climate change science—how it is produced continues to develop and improve.

Climate change is a global phenomenon that has different effects in different regions of the world. Many of these effects will not be fully apparent for decades. This makes it difficult to predict all the impacts climate change will have. Increasingly, scientists are making these projections through climate models— mathematical representations of how human and environmental systems interact.

To construct a climate model, scientists spatially divide each of the earth's components (land, atmosphere, ocean, and ice) into boxes

Doug Clark, Western Washington University (public domain via U.S. National Ice Core Laboratory).

Scientific researchers extract cores of ice from a glacier in British Columbia, Canada in July 2010. Scientists can extract ice samples from thousands of feet below the earth's surface. By analyzing the ice cores, they can study variations in climate going back eight hundred thousand years. Using ice samples, scientists have determined that the amount of CO_2 in the atmosphere was relatively constant until 150 years ago, when it began to rise more than it ever had before. The goal of these studies is to increase our understanding of past climate conditions so we can (1) see how the climate has changed up to the present and (2) develop computer models to predict the impact of climate change in the future.

on a computer. For each box, researchers enter known information about those components and how they interact with the others. They test the model's accuracy by running a simulation into the past and comparing the model's results with observations of what actually happened. Once the model is adjusted to be as accurate as possible, scientists run simulations into the future to make projections about how the earth's climate will change in years to come.

Why do some people doubt climate change science?

Despite an overwhelming consensus among scientists regarding the reality of climate change, some people have expressed doubts about climate change science. They claim that the information from climate models should not be trusted because scientists are just making predictions. Others simply deny climate change's existence without providing any evidence to support their claims.

It is true that since we cannot know precisely what the earth's future environment will be like, there is uncertainty associated with climate models' projections. For instance, we cannot know future greenhouse gas emissions levels exactly; they will depend on many factors including international negotiations, regional political decisions, unforeseen natural events, and technological developments. Despite these uncertainties, the scientific community broadly accepts that climate change will have dramatic effects.

Furthermore, making informed predictions to help plan for the future and manage risk is common in many sectors. Public health officials create plans for disease outbreaks even if there is uncertainty about the likelihood of an epidemic.

The military prepares for many possible conflicts. School administrators plan for a range of potential disruptions—students arriving late, teachers being out sick, and emergency situations like fires and floods.

In each case, decision makers incorporate the best information they have into planning for the future and continue to adapt their course of action as new information becomes available. This is especially important for climate change, where the decisions people make today will affect the environment they live in for decades.

The Effects of Climate Change

Rising temperatures are just one aspect of climate change. The term "global warming" has sometimes been replaced with "global weirding" because there are so many effects of climate change beyond increasing temperatures.

What are the impacts of climate change?

The effects of climate change include rises in sea level, extreme weather events, and threats to human health. Already, many parts of the world are beginning to feel the effects of climate change, while the risk of even greater impacts multiplies each year.

Bre LaRow (CC BY 2.0).

A lobster boat off the coast of Maine in the United States. In recent years, lobster populations have suffered dramatic declines in coastal Massachusetts, Rhode Island, and Connecticut. Scientists attribute the decline in southern New England to rising water temperatures associated with climate change.

Sergey Vladimirov (CC BY 2.0).

On July 29, 2010, temperatures in Moscow, Russia reached 100°F for the first time in the 130 years that measurements have been kept. The record temperature came during a heat wave that lasted for more than three weeks. The heat wave also caused hundreds of forest fires, created smog that blanketed Russia, and contributed to the death of thousands of Russians who were vulnerable to the extreme temperatures and poor air quality. Above, people in Moscow seek relief from the heat by wading in a fountain in a city park.

Oceans: Climate change could raise the level of the world's seas by up to 6.6 feet by 2100. Rising sea levels are caused by polar ice caps melting and by ocean waters expanding as they warm. (As water increases in temperature, it expands to take up a greater volume of space.)

Much of the world's population and many of the planet's most fragile ecosystems could become more vulnerable to coastal flooding. Experts predict that densely-populated coastal cities, such as Calcutta, New York, and Shanghai, could experience more floods. In the southeastern United States, some homes and coastal properties could be underwater within the next thirty years as a result of sea level rise.

Numerous low-lying island countries, such as the Carteret Islands, the Marshall Islands, and Kiribati are becoming engulfed by the sea. Some of their residents, often called the first climate change refugees, have begun to leave the islands. People who lived on the Carteret Islands have already been forced to evacuate, and Kiribati has purchased land from Fiji— over two thousand miles away—so its citizens seeking refuge from sea level rise have somewhere to go.

The ocean also absorbs some of the excess CO_2 in the atmosphere. With more CO_2, oceans become more acidic, which is harmful to marine life and could negatively affect ocean ecosystems for centuries.

> **"Climate change is a challenge that few want to take on. But the price of inaction is so high. Those of us from Oceania are already experiencing it first hand. We've seen waves crashing into our homes and our breadfruit trees wither from the salt and drought. We look at our children and wonder how they will know themselves or their culture should we lose our islands. Climate change affects not only us islanders. It threatens the entire world. To tackle it we need a radical change of course."**
>
> —Kathy Jetnil-Kijiner, poet from the Marshall Islands, in a speech at a UN Climate Summit, September 23, 2014

Extreme weather events: Climate change is affecting weather patterns around the world. In recent years, scientists have observed greater extremes of temperatures (conditions that are either extremely hot or extremely cold), increased numbers of heat waves, and more droughts in many regions of the world.

In addition to temperature extremes, the number and strength of powerful storms has increased. This may be caused by rising ocean temperatures increasing the amount of water that evaporates into the atmosphere. The additional warm water vapor makes storms more powerful. Rising sea levels also increase the amount of damage storms cause, meaning storms that have been less of a problem in the past are now becoming more dangerous. For instance, in May 2017, torrential rains caused flooding along the Mississippi River that led to the deaths of ten people and caused over a billion dollars of damage. It was the latest in a series of increasingly frequent extreme storms to hit the United States in recent years.

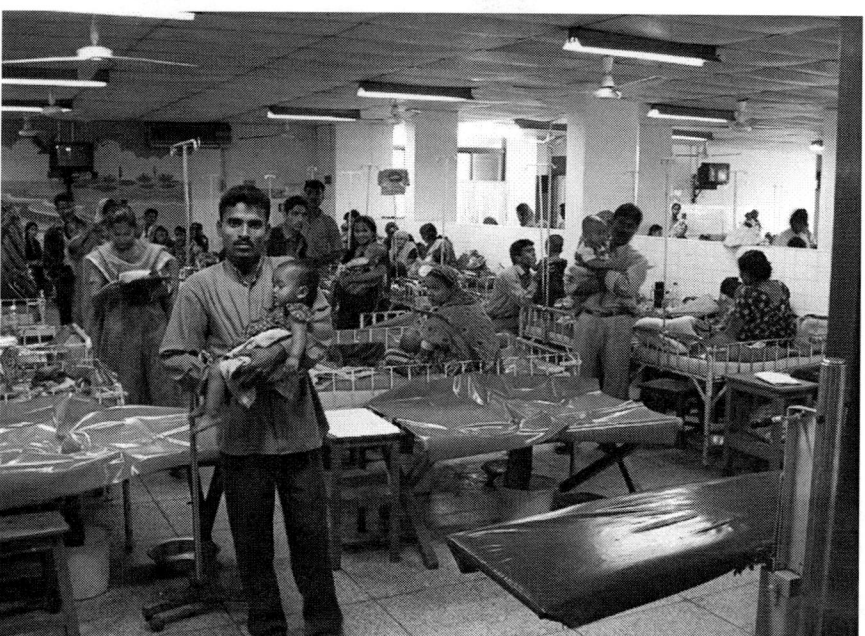

A hospital in Bangladesh for people afflicted with cholera, a waterborne disease. Increases in water temperatures caused by climate change may contribute to the spread of this disease.

Mark Knobil (CC BY-SA 2.0).

> **❝[T]here is a climate change connection, because the oceans and sea surface temperatures are higher now because of climate change, and in general that adds 5 to 10 percent to the precipitation. There have been many so-called 500-year floods along the Mississippi about every five to 10 years since 1993.❞**
> —Kevin Trenberth, climate scientist with the National Center for Atmospheric Research, May 6, 2017

> **❝We're now living in a world of extremes on the Mississippi River. We just don't get normal spring rains anymore. We get huge downpours.❞**
> —Brant Walker, mayor of Alton, Illinois, May 15, 2017

Scientists are generally cautious about saying that climate change caused a particular event. Instead they look for patterns over time and are confident that climate change increases the chance that extreme weather events will occur more frequently.

Health: Climate change impacts human health in many ways. Heat waves and air pollution increase the risk of heat stroke, certain allergies, asthma, kidney disease, and heart disease, especially among elderly people living in urban areas. For example, the 2003 summer heat wave in Europe contributed to over seventy thousand deaths.

In addition, droughts threaten reliable and affordable access to clean water for drinking and food production, which are essential components of good health.

USDA (CC BY 2.0)

In August 2012, an official from the U.S. Department of Agriculture and a farmer inspect a soybean field in Missouri affected by drought. At the time, the Department of Agriculture estimated that about 60 percent of the United States was experiencing extreme drought.

Extreme weather events often have tragic death tolls and destroy the resources needed to promote health among members of affected communities. For instance, powerful storms can restrict transportation, damage medical facilities, and cause power-outages—limiting access to health care. More frequent and intense floods can lead to water contamination and the spread of waterborne diseases like cholera and diarrhoea.

Changes in temperature and rainfall may also shift the geographic ranges of insects that help spread certain diseases. For instance, dengue fever, a mosquito-borne disease of tropical and subtropical regions, returned to Florida in 2009 after being absent for sixty-five years.

Species migration: As drastic and varied environmental changes unfold, some species will be able to adapt to new conditions. Others may have to change their geographic range and relocate to more suitable locations where they have a better chance of survival. Species that cannot adapt or relocate may die out and go extinct.

Plants and animals are generally shifting their habitat ranges either toward the North and South Poles or toward higher altitudes to avoid warming temperatures. Various seasonal aspects of plant and animal life cycles are also being affected, like the migration patterns of birds and insects. For example, moths on Mount Kinabalu in Borneo are flying higher up the mountain (at an average rate of 220 feet in altitude per decade) to escape increasing temperatures.

Many species will not be able to adapt or move fast enough to keep up with the changing climate. Moreover, if species have to move to find new places to live, their migration routes may be blocked by human-made obstacles like dams, roads, and cities or competition with other species. In these cases, the likelihood of species extinction may increase. Each of these effects—species relocations and extinctions—can disrupt entire ecosystems and the valuable services they provide to human societies.

Food and agriculture: The changing climate directly affects food production and could increase costs. With increasing temperatures and changes in rainfall patterns, crop yields in some locations may improve, while in others they may decline. Overall, the negative impacts of climate change will outweigh the positive ones. The yields of major crops like wheat, rice, and corn will decrease in many regions of the world.

More frequent droughts and floods will make food production more difficult for farmers. They may have to completely alter how they approach agriculture in the case of drastic climate change by growing different crops, changing irrigation practices, and using greater quantities of chemical pesticides.

In addition, increasing ocean temperatures associated with climate change will impact fisheries—for example, increasing levels of carbon dioxide in the ocean affect the ability of many fish to thrive. Fisheries and the fishing industry are important to both the food supplies and economies of many countries. Furthermore, these effects will take place at the same time that global demand for food is increasing. These factors could contribute to rising food prices.

Conflict and security: Climate change and its far-reaching environmental effects may also contribute to political conflict and security concerns around the world. In less wealthy countries, where governments are often unable to respond quickly or adequately to disasters, a series of poor harvests or the collapse of fisheries could force millions of refugees across borders. This could lead to violence or governmental collapse.

Alternatively, governments could become more authoritarian in order to deal with these security risks. For example, the Southeast Asian country of Myanmar was devastated by a cyclone (hurricane) in May 2008. Despite more than eighty thousand deaths, the displacement of one million people from their homes, and widespread disease and starvation, the military rulers did not allow humanitarian aid workers to enter the country until weeks after the storm.

Increasingly severe weather systems such as hurricanes, monsoons, or droughts could lead to violence over access to clean water and food supplies. For example, scientists believe that climate change contributes to military conflict in Syria, with warmer, drier conditions leading to increased instances of drought and food scarcity.

> ❝*We're not saying the drought caused the war. We're saying that added to all the other stressors, it helped kick things over the threshold into open conflict. And a drought of that severity was made much more likely by the ongoing human-driven drying of that region.*❞
> —Richard Seager, a climate scientist at Columbia University's Lamont-Doherty Earth Observatory, March 2, 2015, speaking about the effect of climate change on the Syrian Civil War

The U.S. military believes that climate change affects U.S. security. It is also concerned that many U.S. military bases are at sea level and are threatened by the prospect of rising oceans.

> ❝*I agree that the effects of a changing climate—such as increased maritime access to the Arctic, rising sea levels, desertification, among others— impact our security situation. I will ensure that the department continues to be prepared to conduct operations today and in the future, and that we are prepared to address the effects of a changing climate on our threat assessments, resources, and readiness.*❞
> —U.S. Secretary of Defense James Mattis, January 2017, in written testimony to Congress

Conclusion

You have just read a brief overview of the causes and effects of climate change and have seen that a warming world is already influencing the lives of plants, animals, and people across the globe. While understandings of how global warming works and why it is happening have steadily improved over the past few decades, the question of what to do about climate change remains.

In Part II of the reading, you will explore how governments and other groups, including businesses and nongovernmental organizations, are working both to prevent dangerous climate change and to cope with its effects. You will begin to consider who is responsible for the problem of climate change, who is most at risk, and why a unified international strategy for dealing with a changing climate has not yet emerged.

Part II: Responses to Climate Change

The atmosphere is a shared resource, which makes climate change an international problem. To deal with a global problem, policy proposals are made at multiple levels—international, national, and local. This makes agreeing on climate change policies difficult.

Vulnerability and Responsibility

While climate change is a shared concern, it does not affect places and peoples evenly. There are great disputes about who should be held responsible for causing climate change and for repairing its damage.

Who is most vulnerable to the effects of climate change?

Some countries are more vulnerable to the harmful effects of climate change than others. For example, Tuvalu—a country made up of nine small islands in the Pacific Ocean that is home to about ten thousand people—could become uninhabitable in the next fifty years as sea levels continue to rise. Within individual countries, impacts vary by region as well.

Even within local communities, some people may be more vulnerable to the effects of climate change than others. This often depends on where they live and how much their basic needs like food, water, and shelter are affected by changing climate conditions. For example, when Hurricane Sandy hit the United States in 2012, the people who were most affected were those in lower-income communities, such as those living in low-income sections of the Red Hook area in Brooklyn, New York. People in poorer neighborhoods may not have access to health care or the ability to evacuate their homes. Because of this, they tend to be more vulnerable than those who have greater means to recover from disasters.

In many countries, poor communities live in environmentally unsafe areas because this is where they can afford housing. In Nigeria, for example, poor people in the city of Lagos

Part II Definitions

Energy efficiency—Producing more energy with less fuel.

Mitigation—The effort to reduce harm. Climate change mitigation focuses on addressing the causes of climate change. Mitigation efforts seek to reduce greenhouse gas emissions in order to lessen the harmful effects of climate change.

Adaptation—Adjustment to new conditions. Climate change adaptation focuses on adjusting to the effects of climate change. Adaptation efforts seek to reduce people's vulnerability to these effects.

often end up living in swamps or the lowest-lying parts of the coast, where they are most exposed to flooding.

Poverty is important in determining vulnerability to climate change. People in the world's forty-eight poorest countries are five times as likely to die from climate-related disasters. Poor countries are ill-equipped to deal with extreme weather and health issues. Because money is scarce, they often lack effective infrastructure (like hospitals and running water systems) to deal with the impacts of climate change, and their citizens often live in homes that cannot withstand intense storms. Most importantly, countries in the global South are the most likely to already be experiencing issues like water scarcity, food shortages, poor sanitation, and limited access to safe housing. Any worsening of these problems by climate change is likely to be catastrophic.

How have histories of colonialism influenced vulnerability to climate change in the global South?

The economies of many countries in the global South rely heavily on a single agricultural crop. This is a legacy of colonial times, when imperial powers would encourage each

The Global North and the Global South

The "global North" and the "global South" are labels used to differentiate the richer parts of the world from the poorer. The global North includes much of Europe and North America, while the global South refers to large parts of Asia, Africa, and South America. Because most of the richer countries are concentrated toward the north of the globe and most of the poorer countries are toward the south, these geographical labels were adopted. But the terms are not perfect—relatively rich Australia lies in the Southern Hemisphere, while some poorer countries are situated toward the north. For this reason, some people use the terms "developed countries" and "developing countries" instead.

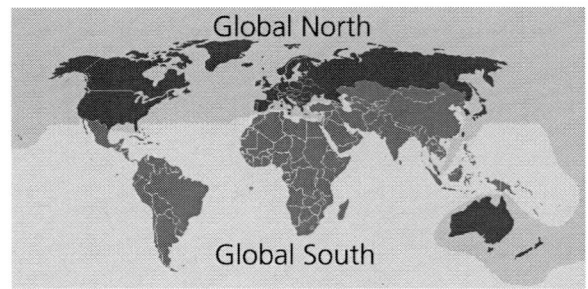

The global North and the global South have political issues with deep historical roots. During the nineteenth and early twentieth centuries, the most powerful countries of what is today the global North competed to establish colonies abroad. Most of these colonies were in the global South. The powerful countries used colonies to gain access to raw materials and to open up new markets for their manufactured goods. They often justified their exploitation of people and resources in the colonies by claiming that they were on missions to "civilize" indigenous communities. By the turn of the twentieth century, Britain, France, and other colonial powers controlled nearly the entire continent of Africa and much of Asia. U.S. colonies included the Philippines and Cuba.

Although almost all of the colonies gained independence by the 1960s, the impact of colonialism has continued to influence international relations. Economic links between the former colonies and the former imperial powers remain important. In addition, leaders in the global South argue that colonialism is the source of many of the problems that currently afflict their countries—from ethnic strife to widespread poverty. International cooperation is complicated by the differences in the priorities that rich and poor countries have and by poor countries' fear that they will be overpowered by the global North. These challenges are apparent in international negotiations about how to respond to climate change, especially when the question arises of who is most vulnerable to its effects and who should be held responsible.

of their colonies to produce only one good or crop. This allowed imperial powers to profit from their colonies while also keeping each colony reliant on the empire for the goods that its people needed. Relying on only one crop makes these countries particularly vulnerable to unpredictable changes in climate and changes in world markets. If the price of their primary good goes down or if they experience one poor crop yield, their entire economy fails, and they experience vast food shortages. This, in turn, makes coping with the effects of climate change an even greater challenge.

How does the question of responsibility affect the response to climate change?

The forty-eight poorest countries in the world, which are home to 12 percent of the world's population and often experience the greatest effects of climate change, are responsible for emitting less than 1 percent of total greenhouse gas emissions. This gap between responsibility and vulnerability complicates any response to climate change and raises important questions:

- Should wealthier countries that emitted more greenhouse gases over the

A slum on the coast of Lagos, Nigeria. Because of devastating poverty, people have been forced to live in areas that face extremely high risks of flooding and damage from storms.

past two hundred years bear more of the responsibility and take on more of the costs of dealing with climate change's effects?

- Should poorer countries that are trying to develop economically cut their greenhouse gas emissions, even if it slows their economic growth?

> **❝Developing countries…are historically least responsible for the emissions that result in climate change, but most vulnerable to its impacts.❞**
> —Jessica Ayers, climate change researcher, "Resolving the Adaptation Paradox: Exploring the Potential for Deliberative Adaptation Policy-Making in Bangladesh," 2011

The dispute over responsibility for past and future emissions is one of the most difficult issues in determining how to respond to climate change. It is costly to respond to climate change. Cutting emissions means putting limits on industry by demanding less use of fossil fuels, which few countries want to do.

What is "climate justice"?

Some activists call for "climate justice" in which countries pay the costs of dealing with climate change proportionately depending on the extent of responsibility for emissions. Those countries that contributed most to the problem would pay the most. Poorer countries, bearing less responsibility for greenhouse gas emissions, would pay less. But it is difficult to motivate wealthier countries to take responsibility, particularly given that they are the least vulnerable to the effects of climate change.

There is also dispute over how responsibility should be calculated. On the one hand, some believe that countries currently emitting the most greenhouses gases should be held responsible. This places much of the burden of responsibility on newly developing countries like China, which emits the most greenhouse gases in the world. On the other hand, China has a large population, which means it does not have the highest per capita (per person) emissions. Newly developing countries like China also do not have long-standing histories of greenhouse gas emissions the way countries like the United States do. Because there are different ways to define responsibility, it is difficult to decide how to create a response focused on "climate justice." This raises another important question:

- Should responsibility be based on current emissions, per capita emissions, or the emissions accumulated over history?

International Responses to Climate Change

In 1992, in Rio de Janeiro, Brazil at what became known as the Earth Summit, 165 governments agreed on the principle of pre-

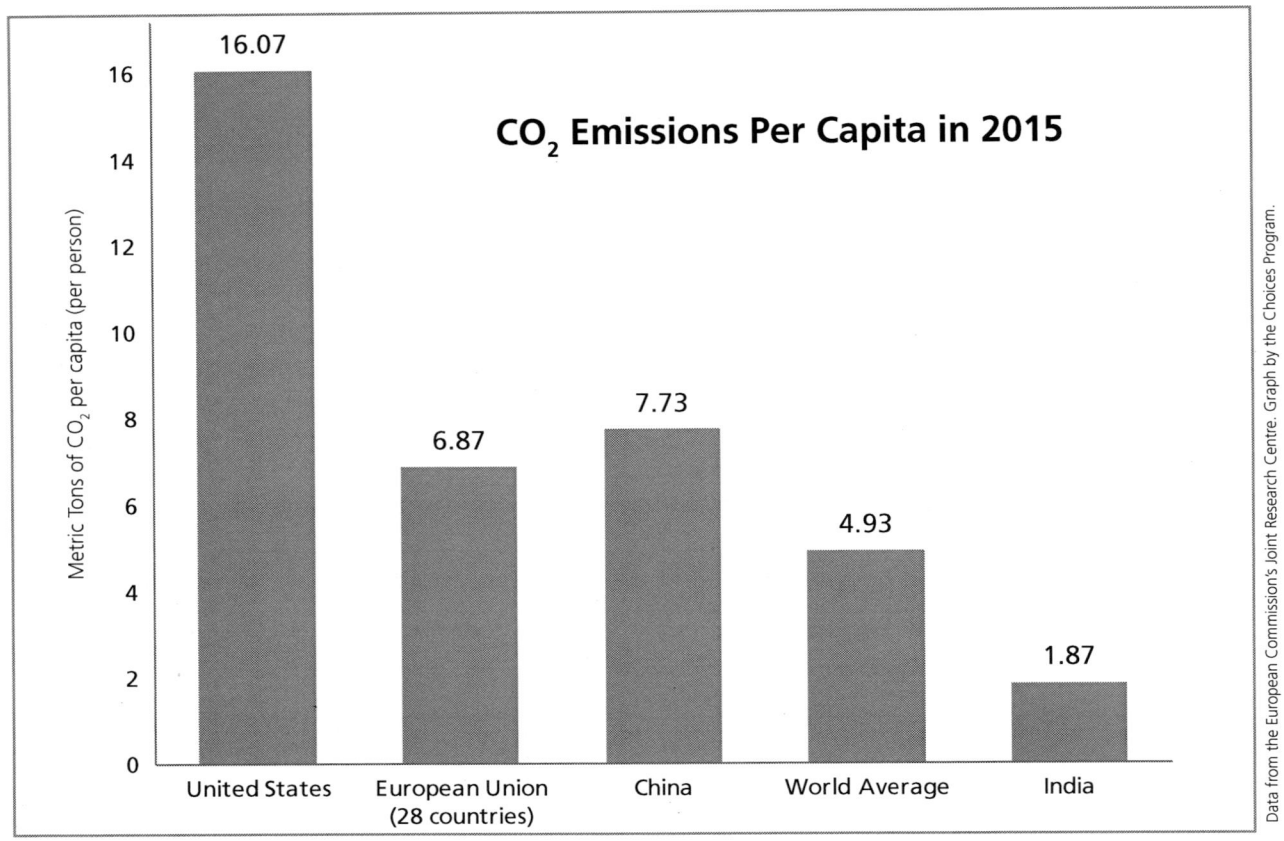

CO$_2$ Emissions Per Capita in 2015

Metric Tons of CO$_2$ per capita (per person)

United States — 16.07
European Union (28 countries) — 6.87
China — 7.73
World Average — 4.93
India — 1.87

Data from the European Commission's Joint Research Centre. Graph by the Choices Program.

This graph shows the average amount of CO$_2$ gas emitted per person in 2015 for the United States, twenty-eight countries in the European Union, China, and India. The world average is also shown.

venting dangerous climate change. In the years that followed, 196 countries ratified the UN Framework Convention on Climate Change (UNFCCC), an agreement to limit greenhouse gas emissions. The agreement stated that while climate change is a shared problem, different countries have different levels of ability to respond, and that wealthier countries should provide funds to address the problem.

Most importantly, the UNFCCC created a system in which countries could continue meeting regularly in order to reach the general goals established in Rio. Each of these yearly meetings is referred to as a "Conference of the Parties" (COP). The COP meetings include government representatives, United Nations (UN) officials, activists, members of NGOs, and corporations that contribute to considerations of how to address the principles agreed to in 1992.

Why have governments struggled to create a unified response to climate change?

It has been difficult for countries to agree on how to limit greenhouse gas emissions and to decide who should make changes to prevent future problems. The economy of a particular country, its peoples' values, and its political structure all contribute to its stance on climate change. For instance, the European Union believes that effective climate change policy must begin with widespread and immediate changes in national and industrial behavior to reduce CO$_2$ emissions. The United States and Japan prefer to focus on developing technology to protect and repair the atmosphere in the future. Poorer countries are primarily concerned with reducing their vulnerability to the effects of climate change.

Despite the principles laid out in the UNFCCC, one of the greatest obstacles in negotiations is deciding who is financially responsible—who should pay. It is initially

A press conference held by Secretary General of the UN, Ban Ki-moon (center) at the 19th Conference of the Parties in Warsaw, Poland. November 19, 2013.

expensive to reduce greenhouse gas emissions because it requires turning away from fossil fuels that are currently the cheapest and most widely used form of energy. But it is also expensive to cope with the effects of climate change.

Determining which course of action to take is particularly tricky because industrialized countries have already reaped the benefits of vast greenhouse gas emissions. This raises important questions:

- Should wealthier countries (historically responsible for the most emissions) be required to contribute money to help poorer countries deal with the effects of climate change?

- Should countries from the global South be subject to restrictions on emissions while they are trying to industrialize?

One of the most significant international agreements, the Kyoto Protocol to the UNFCCC, has fallen victim to these thorny questions.

What is the Kyoto Protocol?

The Kyoto Protocol to the UNFCCC, which was negotiated in 1997, laid out clear emissions restrictions for thirty-seven wealthier countries and reduction targets that poorer countries could volunteer to pursue.

Some countries have viewed the Kyoto Protocol as unfair, because it does not impose restrictions on China or India, both of which are substantial CO_2 emitters. For this reason, in 2007, countries agreed that a new agreement must be drawn up to replace the Kyoto Protocol. In addition, many countries (including Canada and New Zealand) refused to submit to any more emissions restrictions under the existing protocol. The United States did not ratify the protocol at all, making the country exempt from all of the treaty's commitments.

What is the Paris Climate Agreement?

In 2015, representatives from 195 countries went to Paris for the COP 21. Their goal was to create an agreement to replace the Kyoto Protocol (which expires in 2020). Governments from all countries, rich and

Important International Conferences

Where?	When?	What?	Important Developments
Stockholm, Sweden	1972	UN Conference on the Human Environment	Puts the environment on the UN agenda; establishes that it is the responsibility of national governments to protect the environment.
Rio de Janeiro, Brazil	1992	Earth Summit/UN Conference on Environment and Development	Signing of the UNFCCC, which acknowledges the problem of climate change and includes major agreements to stabilize emissions but sets no mandatory limits.
Kyoto, Japan	1997	COP-3 (Conference of the Parties)	Adoption of the Kyoto Protocol to the UNFCCC, which creates binding emissions reduction targets for countries that ratify the protocol.
Marrakech, Morocco	2001	COP-7	Adoption of National Adaptation Programmes of Action (NAPAs) for short-term protection against climate change's effects.
Copenhagen, Denmark	2009	COP-15	Goal of creating a successor to Kyoto is not achieved; Copenhagen Accord (drafted by the United States, China, Brazil, India, and South Africa) only requires that countries pledge to voluntarily reduce emissions.
Cancún, Mexico	2010	COP-16	Many high-emitting countries pledge to voluntarily reduce emissions.
Durban, South Africa	2011	COP-17	Agreement to establish a legally binding deal committing all countries by 2015.
Doha, Qatar	2012	COP-18	Adoption of National Action Plans (NAPs) for long-term protection against climate change's effects; agreement to extend the life of the Kyoto Protocol until 2020.
Lima, Peru	2014	COP-20	In the Lima Accord, all countries (both rich and poor) agree to voluntarily put forward plans to reduce domestic greenhouse gas emissions; there are no requirements regarding the amount of emissions reductions.
Paris, France	2015	COP-21	Adoption of the Paris Climate Agreement, which replaces the Kyoto Protocol. All countries voluntarily pledge to reduce their greenhouse gas emissions. The agreement urges wealthier countries to provide funding to help poorer countries deal with the effects of climate change.

poor alike, agreed for the first time to create voluntarily plans to reduce their domestic greenhouse gas emissions and provide funding for poorer countries' programs to deal with the effects of climate change. The participation and cooperation of the world's largest emitters of greenhouse gases, including China and the United States, helped ensure adoption of the Paris Climate Agreement.

In June 2017, U.S. President Donald Trump announced the United States' intention to withdraw from the agreement. According to the rules of the agreement, a withdrawal could only take effect in 2020, after the next U.S. presidential election. Trump opposed the agreement because he believed it was a bad deal that hurt the economy of the United States. Opponents of Trump's decision included environmental NGOs and large corporations like General Electric, ExxonMobil, and Ford. A poll by the *Washington Post* showed that 59 percent of the U.S. public opposed Trump's decision, while 28 percent supported it.

The light emitting diode (LED) bulb on the left uses over 75 percent less energy than the incandescent light bulb on the right.

What is mitigation?

Responses to climate change are generally categorized into two groups: mitigation and adaptation. The term "mitigation" means efforts to reduce the harm of something. Mitigation of climate change means reducing greenhouse gas emissions with the goal of preventing the harmful effects of climate change.

Many different mitigation strategies have been proposed and used over the past few decades. The production and distribution of many goods and services involves fossil fuels. For instance, plastic bags, water bottles, many plant fertilizers, clothing items, and many cosmetics are all made from oil. Reducing people's demand for these things is one way of decreasing carbon emissions. Increasing energy efficiency (producing more energy with less fuel) across industries is another. Examples of this include using more efficient heating, cooling, and lighting systems in buildings.

Industry standards and regulations can be used to promote mitigation efforts. For instance, requiring certain levels of fuel efficiency for cars or changing building codes can decrease greenhouse gas emissions throughout an industry. Adding labels to products with information about how they were produced can also reduce emissions by educating consumers and perhaps changing what consumers choose to buy.

Another strategy involves reducing the emissions intensity of fuel sources. This means switching from fuels like coal and oil, which emit a lot of CO_2 when used, to fuels like natural gas, which emit less.

A wind farm outside of Copenhagen, Denmark, which supplies 4 percent of the power for the city. Wind spins the blades of the wind turbines, which connect to generators inside the turbines that create electricity.

Furthermore, increasing the use of zero-emissions energy sources is important in reducing the amount of greenhouse gases that end up in the atmosphere. Nuclear energy as well as energy produced from renewable sources like solar, wind, and water power do not emit any CO_2 when used.

But zero-emissions energy sources come with their own set of problems. There are safety concerns about nuclear energy production and the storage of its radioactive waste. There is also a question of whether renewable energy can be produced at a low enough cost and a large enough scale to be widely used. Both businesses and people can be reluctant to pay more for zero-emissions energy if they can pay less for energy from fossil fuels.

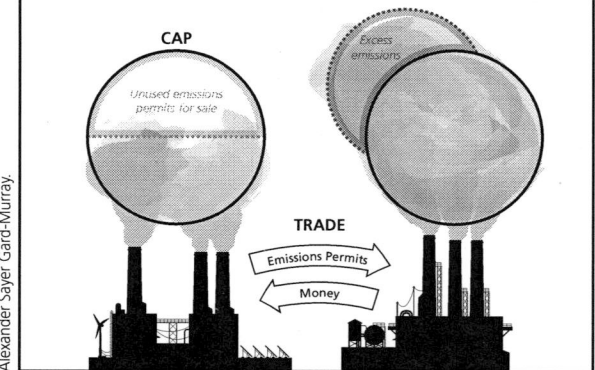

A cap-and-trade system sets a limit to the amount of greenhouse gases companies can emit. It allows companies that want to emit more than their share to buy emissions permits, or allowances, from other companies that emit less.

What economic policies encourage mitigation?

Different types of policies encourage the use of these mitigation strategies. Several countries have implemented carbon taxes, which require emitters to pay fees for the amount of CO_2 they emit. The money generated from the taxes can then be used to invest in other things, like the development of renewable energy.

In cap-and-trade systems, a limit, or "cap," to the total amount of emissions is set. Permits that allow companies to emit certain amounts of greenhouse gases are then either given away or auctioned off. If one company wants to emit more CO_2 than it has permits for, it can buy permits from another company. Likewise, if a company does not need to use all of its permits, it can sell them to other companies. Both carbon taxes and cap-and-trade strategies attempt to put a price on carbon. If the price is too high, the economies of participating countries will suffer. If the price is not high enough, emissions levels will not drop significantly and may even rise in the long term.

What is adaptation?

While mitigation efforts are aimed at reducing the amount of greenhouse gases in the atmosphere, adaptation focuses on adjusting to the effects of climate change.

Geoengineering

A few people have proposed some more extreme ways of fighting the effects of climate change: firing chemicals into clouds to change rainfall patterns, shooting pollution into the atmosphere to block sunlight coming toward the earth, and using machines to suck CO_2 out of the atmosphere and burying it underground. These strategies, collectively known as "geoengineering," use large-scale technological fixes to drastically change the planet's environment. While they attempt to reduce or slow some of the effects of global warming, these strategies are not aimed at addressing climate change's underlying causes. There are many serious safety risks associated with geoengineering, as well as the potential for unknown side effects—especially because there is no way to test such large-scale projects before implementing them. Furthermore, these strategies are expensive, and only the wealthiest countries would be able to afford them. With countries in the global North in control of geoengineering efforts, the interests of countries in the global South could be lost.

In the decades following the 1992 Earth Summit, international leaders primarily directed their attention to mitigation. They hoped that through decreasing CO_2 emissions, many of the potential negative effects of climate change could be avoided. Disagreements among countries about how to actually implement mitigation strategies have meant that progress on reducing greenhouse gas levels has been slow. Efforts to reduce emissions continue, but even if all human emissions of CO_2 somehow stopped today, emissions from the past that have accumulated in the environment would still cause continued climate change. Scientists and policy makers have now recognized that mitigation alone is not enough.

The Cancún COP in 2010 declared that adaptation must be equal in priority to mitigation. This means that greater attention is now being paid to helping countries and communities adapt to the effects of climate change.

Adaptation strategies vary. Urban planners in coastal cities, for example, may take into account sea level rise and flood surges that accompany extreme storms. Farmers may plant different crops that are more resilient after droughts and floods. Governments might implement early warning systems for more frequent and intense extreme weather events and disease outbreaks.

Improved access to health care and economic opportunity are also important in re-ducing people's vulnerability to the effects of climate change. For instance, creating new types of jobs in communities that have been dependent on fishing will make the problem of shrinking fish populations (caused by climate-related changes in ocean conditions) less catastrophic. People will have other options for how to make a living. Adaptation needs and priorities vary significantly depending on the environmental, social, and economic conditions of different countries, regions, and individual communities.

What are National Adaptation Programmes of Action?

The increasing emphasis on adaptation, and the fact that adaptation needs are specific to particular places and peoples, has led to the development of National Adaptation Programmes of Action (NAPAs). The world's poorest and most vulnerable countries have submitted NAPAs, which are plans that outline their most urgent needs regarding climate change adaptation, to the UNFCCC. Each NAPA consists of a ranked list of projects the country has identified as its highest priorities in adapting to climate change. The UNFCCC has a fund provided by wealthier countries to help implement those plans. As of 2015, about $320 million in funds have been distributed, although this is not nearly enough to fund all of the NAPAs.

NAPAs, adopted at the COP in Marrakech, Morocco in 2001, focus on countries' most immediate adaptation needs. Later, in Doha, Qatar in 2012, National Adaptation Plans (NAPs) were adopted by the COP to address more medium- and long-term projects to adapt to a changing climate.

While decisions about the global response to climate change are made at the international level, NAPAs and NAPs depend on extensive input from local stakeholders (people and organizations who have an interest in or are affected by an issue). Both NAPAs and NAPs are intended as ways for members of local governments and communities to participate in making decisions about how their country will adapt to climate change. They are structured this way so that adaptation plans are not imposed from above, but rather driven by the concerns and priorities of the people in each country. For although climate change poses risks on an international scale, many impacts of climate change are experienced locally.

Practical Action Bangladesh. Used with permission.

People in Bangladesh, one of the most vulnerable countries to sea level rise, have adapted in some places by creating farms that float on water.

> ***The lesson from climate change is... risks do not register their effects in the abstract; they occur in particular regions and places, to particular peoples, and to specific ecosystems.**"*
> —Professors Jeanne X. Kasperson and Roger Kasperson, 2001

Finding a way for local communities to participate in decision-making about climate change adaptation on national and international levels can be challenging. This is because people and groups with more power, such as wealthy countries and international organiza-

Paying for Adaptation

Many challenges and questions arise in determining how to pay for adaptation to climate change. Mitigation efforts, such as emissions reductions, have positive effects that are felt around the world, whereas adaptation tends to have more local benefits. As a result, mitigation has been more appealing for wealthy countries to fund because they will directly experience its positive effects. Consequently, much more money overall has gone toward mitigation efforts than toward adaptation.

In addition, while it is generally agreed upon that money to help poorer countries cope with climate change's effects should go to the most vulnerable countries first, which countries and projects should get prioritized is hard to determine. This is especially challenging when there is a limited amount of money available and contributions from wealthy countries are voluntary. At the 15th COP in Copenhagen, Denmark, wealthy countries promised roughly $50 billion per year to go toward adaptation by 2020. Some experts believe that a great deal more will be needed to fund adaptation projects.

tions, often see local communities' expertise as inferior. For example, the top priority project in Bangladesh's NAPA was to establish more forests in coastal areas to protect against sea level rise and more frequent extreme weather events. The NAPA process included input from people living in coastal communities, but their participation was mostly sought out to confirm what higher-up officials and other experts had already decided were the top concerns. As a result, this project focused on dealing with the physical effects of climate change, while the local stakeholders wanted instead to focus on improving social and economic conditions so they would be less vulnerable. In this case, the priorities reflected in the NAPA did not match those of the local stakeholders.

What is sustainable development?

Sustainable development is a way of using resources that protects both environmental and human well-being in the long term. Responding to climate change—through both mitigation and adaptation—can present challenges for countries worried about it hurting their economic growth. Acknowledging this concern, international agreements and conferences have stressed the importance of sustainable development to combine economic improvement and climate change prevention. The goal of sustainable development is to meet the economic and social needs of the present generation without compromising or depleting resources for future generations.

In the case of climate change, this principle is particularly important as the most severe effects of today's actions will most likely only be seen in decades to come. Fossil fuels, in particular, are not a sustainable source of energy. In addition to the harmful effects of the greenhouse gases released when people use them, the supply of coal, oil, and natural gas is limited. These energy sources take a long time to form, for they are made from the remains of plants and animals that lived millions of years ago. People are using them up at a faster rate than they are being regenerated, which makes humans' current use of fossil fuels unsustain-

able over a long period of time. Furthermore, as they become less easily available, they will also become more expensive.

Why do countries pursue sustainable development?

Sustainable development is a principle that many rich and poor countries try to follow. Some European countries, for instance, promote the use of bicycles to reduce reliance on fossil fuels. Efforts like this show that, in the global North, resources can be used sustainably without sacrificing a high standard of living.

In the global South, renewable energy is an important aspect of sustainable development, providing struggling communities with effective and efficient power sources. For example, some regions of the Philippines use solar power to pump and purify drinking water. Remote villages in northern Peru generate electricity using the high levels of rainfall they experience. These sustainable development efforts both reduce the emission of greenhouse gases and provide employment opportunities.

Many policy makers and economists see sustainable development as a huge opportunity for the global South. They suggest that poorer countries can "grow green" by using technology that was not available when richer countries first industrialized. Because of this, sustainability is often included in the conditions of loans offered by wealthy countries to the governments of the global South. Sustainable development also has the potential to help many poorer countries become less reliant on foreign aid.

National and Local Responses to Climate Change

Frustration with the slow pace of the UNFCCC process to combat climate change has motivated other people and groups to propose their own responses. This has happened at different levels, from national governments to small interest groups. For example, nearly five hundred presidents of colleges and universities in the United States have vowed that their institutions will become carbon neutral (contributing no additional CO_2 into the atmosphere). In many parts of the world, national and local governments are passing their own rules about climate change. Organizations, businesses, and individuals are also devising responses to the climate change.

National governments: Some countries are opting to take a dramatic stance on climate change despite the lack of legally binding international agreements. Denmark has vowed to end all fossil fuel use by 2050 and is a leading user of energy from wind turbines. Costa Rica, aiming to be carbon neutral by 2085, imposed a carbon tax on fossil fuels. Part of the money from this tax goes toward forest conservation.

The United States, though it is the largest historical emitter, has demonstrated less enthusiasm. In November 2014 under President Obama, the United States announced a new goal of reducing its emissions 26 to 28 percent below 2005 levels by 2025. But in June 2017, President Trump announced that the United States would withdraw from the Paris Climate Agreement and would no longer attempt to meet the emissions target of a 26 to 28 percent reduction.

Local governments: Although the United States has been slow to take action on climate change at a national level, some individual states have not shied away from creating climate change policy. For instance, Massachusetts became the first state to place limits on greenhouse gas emissions from power plants in 2001. This led to the establishment of the country's first cap-and-trade program, the Regional Greenhouse Gas Initiative, involving nine states along the East Coast of the United States.

In 2006, California passed the Global Warming Solutions Act, which requires emissions be reduced by nearly 30 percent statewide by 2020. Individual cities and towns are also making climate change a policy priority. Immediately after President Trump announced the United States would withdraw from the Paris Climate Agreement, the governors of twelve states and more than two hundred mayors announced plans to continue to try to meet the 2025 reduced emissions target.

Businesses: Governments are not the only institutions that have responded to concerns about climate change. Even businesses, long seen as the enemies of climate change policy, have become increasingly invested in reducing their impact on the climate.

When the Kyoto Protocol was signed, oil producers, vehicle manufacturers, and electrical trade associations vowed to prevent the ratification of the agreement. Because of their economic strength, they have had the political power to stall many mitigation policies. However, in September 1997, the oil corporation BP announced that it would voluntarily measure its emissions and research how to reduce levels of greenhouse gases.

By acknowledging the effect of fossil fuels on the environment, BP set off a revolution in how corporations approached climate change. Companies and industry representatives began to actively participate in the COP meetings and presented formal reports on their own attempts to fight climate change. Businesses began investing in forestry

Henrik Boegh (CC BY-SA 3.0).

Public bicycles and abundant bike lanes are part of Denmark's policy to reduce CO_2 emissions from cars by making Copenhagen a cycling city.

Campaigners from Friends of the Earth International participate in a peaceful march to demand climate justice during UN talks in Copenhagen. December 2009.

projects, which aimed to provide more CO_2-absorbing plants, and started promoting the idea of emissions trading.

Some people claim that these companies are only trying to appear more environmentally friendly to improve their public image. For instance, while BP advertises its strong commitment to renewable energy, nongovernmental organizations such as Greenpeace have drawn attention to the fact that the company invests significantly less money in alternative energy sources compared with its investments in oil and gas. In addition, ExxonMobil, one of the world's largest oil and gas companies, prominently advertises its commitment to reducing the risk of climate change. But at the same time, the company provides millions of dollars each year to organizations that promote the denial of human-caused climate change and to campaigns for politicians who are against climate change mitigation policies.

It remains true that corporations lobby against policies like carbon taxes and that the amount of money oil companies put toward alternative energy is much less than what they spend to increase oil extraction. However, even the fact that businesses are feeling pressure to present a new, green face and that customers are increasingly drawn to companies that express concern about the environment is significant.

The media: For a long time, media coverage of climate change, especially in the United States, attempted to balance claims of global warming with counter-claims that climate change was not real or that human activity was not causing it. This has declined as overwhelming scientific evidence of human-caused climate change has emerged.

Media coverage of climate concerns has ballooned in recent years. This means people around the world have more access to information about climate science and how the atmosphere is changing. It also means that there is greater awareness of climate-related incidents such as severe storms. Because of the media's growing emphasis on climate change, populations around the world have become more aware of the dangers it poses to them and others.

Nongovernmental organizations (NGOs): Putting pressure on government is one of the key functions of "green groups," environmental advocacy organizations such as Friends

The Motion is:
"We save the planet,
but not yet..."

POZNAN
Climate Change
Conference

of the Earth and Greenpeace. Like the media, these organizations are responsible for increasing public awareness of climate change. They frequently engage with people about their personal greenhouse gas emissions, as well as launching organized efforts to confront politicians. Other types of NGOs focus on providing on-the-ground assistance to communities trying to decrease their vulnerability or in climate change-affected areas that need disaster relief. Examples of these include Oxfam, Christian Aid, and the International Federation of Red Cross and Red Crescent Societies.

The influence of NGOs is becoming increasingly central to climate change negotiations on national and international levels. The number of organizations authorized to participate in COP meetings has expanded considerably over the years, including many from both the global North and the global South. In fact, at some COPs, representatives from NGOs have outnumbered government negotiators. At these meetings, the organizations take part in negotiation sessions, distribute reports, hold events, and interact with the press.

Growth in the number and diversity of organizations attending negotiations has led to greater influence for NGOs. But there are also divisions within the NGO community. These organizations do not have uniform priori-

ties and often disagree on things like the viability of alternative energy or how to prioritize between mitigation and adaptation. NGOs, therefore, do not present a wholly unified force during negotiations.

Barriers to Action on Climate Change

Global climate change has been on the agenda of local governments, international leaders, and nonprofit organizations for decades, yet sometimes it can seem like little has been accomplished. Many significant obstacles stand in the way of responding to climate change.

What are the most significant obstacles to responding to climate change?

Cost: Reducing greenhouse gas emissions can be a costly endeavor. Right now, fossil fuels are the cheapest source of energy. This is largely because of benefits called subsidies that governments provide. Subsidies make energy from fossil fuels cheaper to produce, lower the price consumers have to pay for that energy, and increase the price energy producers receive. Furthermore, fossil fuels are used in manufacturing goods, in generating electricity, and in other industrial activities central to a country's economic growth. In comparison, other sources of energy, such as wind farms or nuclear power plants, are expensive to build.

Mitigation policies like carbon taxing or cap-and-trade systems attempt to make fossil fuels more expensive so that companies will find alternative energy more attractive. Unfortunately, this means that the price of products would increase, and neither businesses nor individual people want to pay more. If goods in one country increase in price, that country is at a disadvantage when trying to sell its products in global markets.

Many experts say that for mitigation to work, societies and cultures have to change in order to use less energy. This requires a transformation in how people live their lives, challenging the way homes are constructed, the kinds of food available to eat, and even the idea of individuals owning cars. Many people see this as a cost in quality of life. Ultimately, any sort of change will generate resistance. Even reducing a country's vulnerability through adaptation, often seen as the "cheap" version of climate change policy because it does not scale down industry, demands funding. Obtaining this funding can be a serious challenge to governments and other organizations.

> ❝[I]t's always easier to shell out money for a disaster that has already happened, with clearly identifiable victims, than to invest money in protecting against something...in the future.❞
>
> —James Surowiecki, a journalist in the *New Yorker*, 2012

The North-South divide: The historical tensions between rich and poor countries can complicate the processes of climate change negotiation. While individual countries have different priorities and interests, poorer countries often feel they are not being treated fairly. Countries from the global South object when mitigation proposals restrict their ability to grow economically, making it harder for them to build roads, provide electricity, improve education, and create jobs for their citizens. They want the same opportunities that wealthier countries have had to pursue prosperity.

Many poorer countries are also rightfully concerned about being overpowered in negotiations. The global North has more bargaining power in the international system—rich countries have the money to control world markets, which means they can control the economies of other countries. Wealthier countries often use their power to pursue their own national interests. Poorer countries, on the other hand, do not have the economic influence to push

other countries to compromise. The result is that the most vulnerable countries have the least success including their interests in international agreements.

Most importantly, countries of the global South struggle to afford mitigation and adaptation projects, despite the fact that they will be most intensely affected by climate change. The countries that can afford these policies do not face such immediate or hard-hitting effects and therefore find the problem of climate change less urgent.

Political disagreement: Competition for political power often makes responding to climate change difficult. At the international level, government officials can be pressured by leaders from other countries, by economic concerns, and by groups of people within their own country who have strong and specific interests. These competing demands mean that even though the risks associated with climate change are clear, some leaders may end up prioritizing other issues in order to maintain relationships and power.

In addition, whatever agreements are accomplished at the international level ultimately come back to individual countries to carry out. For this reason, political conflicts at national and local levels also help determine what action people take to mitigate and adapt to climate change.

For example, there is strong partisan disagreement among politicians in the United States about climate change. The disputes include concerns about the impact of climate change mitigation on jobs and businesses, what role the international community should play in making policy, and even debates about climate science.

Furthermore, in many democratic countries, government leaders are elected every few years. It can be difficult for a country to make lasting decisions about climate change if its primary leaders change so frequently. Also, with politicians regularly up for re-election, they may choose to focus on issues that provide short-term benefits to the people who might vote for them as opposed to focusing on mitigation and adaptation strategies that are initially expensive but reduce longer-term risks. This way, a politician is more likely to retain his or her power and influence, even if the risk of dangerous future climate conditions continues to grow.

In addition, certain industries have a stronger presence in some countries than in others. Some of these industries (like renewable energy and agriculture) may benefit from responding to climate change while others (like oil and coal production) will not. This second category of industries can put immense pressure on government officials to make decisions that support the continued use of fossil fuels. Corporations in these industries do this through lobbying, providing funding for political campaigns that align with their interests, threatening to withdraw financial support from individuals or institutions if certain political decisions are made, and carefully crafting advertisements and publicity campaigns.

Communication: Climate change is a tricky issue for scientists, journalists, policy makers, and the general public to talk about. The concept of climate change can be hard to fully grasp because it refers to changes that occur over long periods of time. Because global warming's effects may not be visible from one day to the next, climate change is less easily relatable to people's daily lives and can be easy to ignore or put off until later.

> **❝You will never see a headline that says 'Climate change broke out today.'"**
> —Andrew Revkin, *New York Times* reporter, 2007

Furthermore, scientists who study climate change think about all the intricate details involved in the systems they study and often use highly specific terminology. As a result, they sometimes struggle to express to the public what the one or two main take-away points of their research are and why their findings matter.

Scientists are also trained to emphasize what they do not yet know and to make all potential uncertainties very clear. Government officials and journalists generally want to hear what scientists do know so that these discoveries can help inform important policy decisions. This tension in communication style often makes scientific conclusions about climate change appear less confident in the media than they really are. Scientists' careful explanations of uncertainty get misinterpreted as meaning that they are not sure of their findings. This may be one reason why the broad scientific consensus about the dangers of climate change has been misrepresented.

Finally, climate change is often talked about as having potentially catastrophic effects. Emphasizing the dangerous impacts of climate change can make the issue feel overwhelming and hopeless. If people think that there is no possible solution to climate change, they may not be motivated to take action to slow or stop its effects. Thinking carefully about how people talk about climate change and making sure to highlight how much can be done is crucial to keeping people engaged with the issue.

Conclusion

You have read about the causes and effects of climate change, tracing its impacts on areas as varied as health, species migration, agriculture, and international security. You have explored contested understandings of who is responsible for global climate change, who is most vulnerable to its effects, and who has the ability to respond. You have also thought through different types of responses—mitigation and adaptation—weighing the benefits and barriers associated with each.

Climate change is a global issue with locally felt effects. While you have read about the complex web of international climate change conferences and the variety of local actors involved, you will next dive into case studies about how specific communities around the world are experiencing and responding to climate change.

Part III: Case Studies

The case studies you are about to read are not meant to be exhaustive or comprehensive. They are designed to highlight a range of effects climate change has on people and places around the world, to consider what factors in specific situations shape the effects of climate change, to think about how these factors interact to increase vulnerability of specific groups of people, and to explore various methods of response.

Each case study includes a table with basic information about the country covered. For those that focus on cities or states, information about the specific regions is included in the text of the case. You will be given the population of each country and its gross domestic product (GDP) per capita. GDP per capita is an estimate of the economic output per person within a country and gives a sense of a country's economy and its citizens' incomes. A higher GDP per capita generally means the population of that country is more wealthy and the government has more money for pursuing various policies. A lower GDP per capita, on the other hand, suggests a poorer population with limited government funds.

In addition to population and GDP per capita, the table for each case will show the average person's life expectancy and the carbon dioxide (CO_2) emissions per capita (per person) in that country. Use the information

> ## Part III Definitions
> **Gross domestic product (GDP)**—The total value of goods and services produced within a country in a year.
>
> **Life expectancy**—The average number of years that a person can expect to live. It is based on statistical analysis of social, economic, and biological data.

in these tables as you consider each case and compare it with others. As you read, keep these questions in mind:

- What might be the priorities of the people living in each country?

- To what degree is each country responsible for the greenhouse gas emissions that are warming the planet?

- Which effects of climate change might be of greatest concern to each country?

- To what extent can each country respond to the effects of climate change?

- Who are the people driving efforts to respond in each place?

After reading these case studies, you will begin to consider what responses to a changing climate, both locally and internationally, you believe to be most fair and effective.

California, United States

The story of California's 2006 Global Warming Solutions Act provides a window into climate change mitigation politics.

With over thirty-nine million people, California has the highest population of any U.S. state. If it were a country, its economy would be sixth largest in the world. Both of these factors contribute to California's ability to influence the rest of the United States, helping it set policy trends that the rest of the country may soon follow. Californians have a history of being interested in environmental issues. This public support has helped the state lower its emission levels. (California emits 9.2 metric tons of CO_2 per capita, while the U.S. average is about 16.2 metric tons.) Looking more closely at California can help show how mitigation policies can be established locally rather than at the international level.

United States	
Population	324 million
GDP per Capita	US$57,300
Life Expectancy	80 years
CO_2 Emissions per Capita	16.2 metric tons

How is California experiencing climate change?

Jerry Brown, who became governor of California in 2011, has called California the "epicenter of climate change." The state is experiencing a number of climate change's effects—flooding, lower crop yields, extreme heat, drought, and wildfires—which are expected to worsen in the coming century.

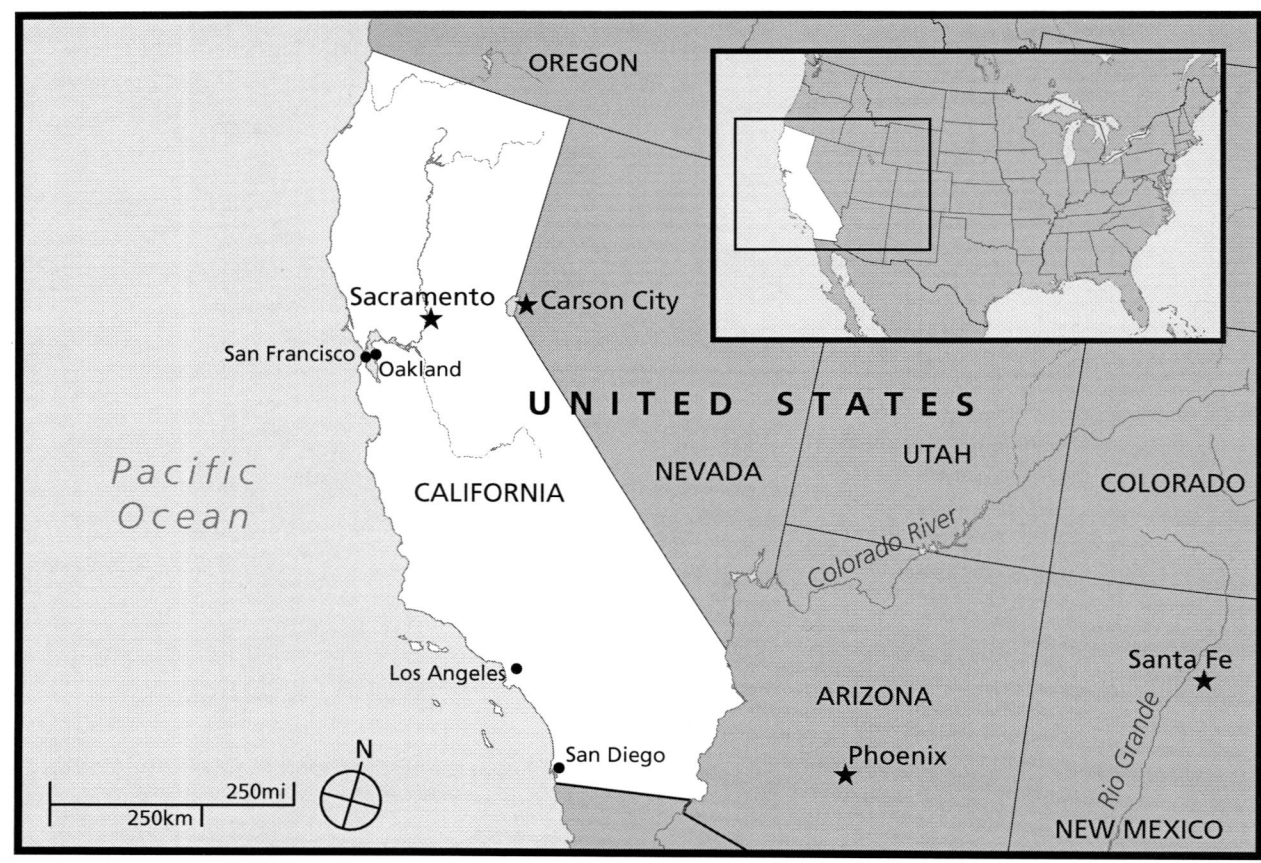

California's heavily populated coastal cities are highly vulnerable to climate change. Many of these cities are important shipping hubs that bring critical goods to the United States. Sea level rise associated with global warming will increase the risk of flooding and damage to highways, power plants, airports, and other infrastructure along California's coast. California also produces over a third of the United States' vegetables and two-thirds of the country's fruits and nuts. Changes in water supply and temperature related to climate change will reduce crop yields in the state and affect the many farming communities that depend on agriculture for work.

In addition, California is already prone to drought, and climate change may make future droughts more intense. Political conflicts have erupted within the state as well as between California and its neighbors about how scarce and precious water resources should be shared. An extreme drought has already caused tens of thousands of farm workers to lose their jobs and has cost the agriculture industry well over a billion dollars. Hot, dry conditions also increase the risk of wildfires,

which can destroy people's homes. Some experts predict that the area of land in California affected by wildfires will increase significantly in the next hundred years.

How is California responding to climate change?

California has taken the lead on climate change mitigation within the United States. In 2006, California passed a law called the Global Warming Solutions Act, which requires that by 2020, the state reduce its greenhouse gas emissions to the levels they were in 1990. If California meets this goal, its 2020 emissions would be nearly 30 percent less than they would have been without the law, bringing the state close to what the Kyoto Protocol would have required.

The law also includes plans for a statewide cap-and-trade system for major emitters, one of the first in the United States. This program is seen as a "test case" as to whether a cap-and-trade system can successfully curb emissions without hurting the

Climate change is affecting agriculture in California, which grows much of the United States' fruits and vegetables. Water shortages from more intense droughts could threaten farmers' ability to water their crops.

economy. Environmental regulations in California have influenced national policies in the past. If the program is successful, some say it will increase the chance that the United States could adopt a similar program nationally.

Establishing California's Global Warming Solutions Act was not easy. Even after the law was passed, it was nearly overturned. Oil companies, even those from other states, funded much of the opposition to the law. These companies claimed their disapproval was fueled by concern that the law would cause many Californians to lose their jobs. But many others, including California's governor at the time, believed the companies were worried that their own profits would suffer under the law.

U.S. Forest Service Chief Tom Tidwell (left) and Smokey Bear (right) honoring then Governor of California Arnold Schwarzenegger (center) for signing and implementing California's Global Warming Solutions Act of 2006.

> **"Does anyone really believe that these companies, out of the goodness of their black oil hearts, are spending millions and millions of dollars to protect jobs?... It's not about jobs at all, ladies and gentlemen. It is about their ability to pollute and thus protect their profits."**
>
> —Former Governor of California Arnold Schwarzenegger, 2010

Oil companies were not the only groups that resisted. Some environmental groups were concerned that the cap-and-trade system did not put strong enough limits on emissions. They feared that it would be ineffective in mitigating climate change and that it would not do enough to stop oil companies from polluting areas where many poor people live. Despite so much controversy, the law remained in effect, and California implemented its cap-and-trade system in 2013.

By 2020, experts say the state's cap-and-trade program will be generating $5 billion per year (as companies purchase emission permits). California plans to put this money toward other climate change mitigation and adaptation projects, and at least 25 percent of the money will go toward programs that benefit poor communities. The state has also partnered with the Canadian province of Quebec—linking their cap-and-trade systems together—and plans to continue joining with other programs.

California continues to push forward in climate change mitigation. In 2015, Governor Brown announced a new goal for the state, calling for 50 percent of its energy use to come from renewable sources by 2030. The following year, California enacted legislation to reduce greenhouse gas emissions to 40 percent below 1990 levels by the year 2030. Even though California is not a national government, the state is providing a model for mitigation on a national and international scale.

China

China's recent and rapid development demonstrates how economics influences both vulnerability to and responsibility for climate change.

The People's Republic of China is the world's largest country by population size and the second largest by land area. Within the country there is huge variation in culture, living standards, and environmental conditions across provinces.

For more than thirty years, China's economy has been steadily growing, which has drastically reduced poverty. At the same time, there has been a rise in inequality between China's rich and the country's poor. The case of China illustrates the impact of economics on both vulnerability to and responsibility for climate change.

China	
Population	1.4 billion
GDP per Capita	US$14,600
Life Expectancy	76 years
CO_2 Emissions per Capita	7.5 metric tons

How is China experiencing climate change?

The impacts of climate change in China are often overlooked in the local media and in political discussions. Nevertheless, climate change affects China in many important ways; the country faces risks of flooding, extreme weather, and food and water scarcity. Some scholars have even claimed that China is the country most vulnerable to the effects of

This map shows China's population density (how many people live in each square kilometer of territory). China has the highest total population of any country in the world, and much of its population is concentrated along the coast.

climate change. China's meteorological administration has said that in parts of the country, the speeds at which temperatures are rising have consistently topped the global average.

Dramatic temperature increases mean that people in China can expect to experience severe social effects of climate change. Some of the most densely populated cities in the world are situated along China's coast. The city of Shanghai, for example, has a population density of more than 9,900 people per square mile (far exceeding Los Angeles' population density of 8,200 people per square mile). In addition, the rural areas surrounding these coastal cities have much larger populations than inland areas toward the west. Because of the concentration of people living along the coast, sea level rise threatens the homes of eighty-five million people (more than the entire population of the United Kingdom).

China's strict residential laws (known as the *hukou* system) that limit the ability of people to migrate could make this even more devastating. In particular, it is very difficult to move from rural areas in the countryside into a city. Rural residents are much poorer than those who live in the cities. Many of these rural economies offer few employment options other than farming. As a result, people in rural areas are hard-hit by crop failures caused by extreme weather or changes in the lengths of seasons. Additionally, a lack of hospitals and other services and infrastructure means that rural residents find it extremely difficult to recover from the effects of climate change. These factors, combined with how difficult it is for people to leave rural areas, make China's rural population especially vulnerable to the effects of climate change.

How is China responding to climate change?

Despite China's vulnerability to many dangerous effects of climate change, the Chinese government has been reluctant to commit to strong mitigation agreements. This is because China's economy is dependent on manufacturing that requires a great deal of energy and that relies largely on coal. Until the rapid economic growth of the past thirty years, China was a very poor country. It has used the rapid expansion of industries to relieve poverty. Promising to sharply reduce emissions from fossil fuels might slow China's economic growth.

China's historical reluctance to increase its mitigation efforts have had international consequences. For example, the main reason that the United States refused to ratify the Kyoto Protocol was that the treaty did not require emissions reductions from countries in the global South, like China and India.

Frog and Onion (CC BY 2.0).

A farmer in the Yunnan province of China.

> **We still have 150 million people living below the poverty line and we therefore face the arduous task of developing the economy and improving people's livelihood. China is now at an important stage of accelerated industrialization and... we are confronted with special difficulty in emission reduction.**
> —Former Chinese Premier Wen Jiabao, at the Copenhagen Climate Summit, 2009

In fact, China currently emits the most greenhouse gases of any country, accounting for about 30 percent of the world's CO_2 emissions. This is partially because it has a large population; the amount of greenhouse gas emissions per person in China is actually less than half that in the United States.

China's status as the "factory of the world" also contributes to the country's high emissions levels. Many of the products used across the world are manufactured in China, so some claim that emissions have been "exported" to China. This means that despite China's position as the biggest emitter, it may not be the most responsible for climate change because it is merely emitting gases on behalf of the countries that are buying its goods.

But China is starting to show greater interest in reducing its dependence on fossil fuels. In November of 2014, Chinese President Xi Jinping announced a joint emissions reduction agreement with U.S. President Obama. This was China's first ever commitment to cap its CO_2 emissions and paved the way to its agreeing to sign the Paris Climate Agreement in 2015.

In addition, China is showing an increasing interest in alternative energy. The country has become the world's leading investor in renewable energy technology both at home and abroad. China now produces more wind turbines and solar panels than any other country. China wants to be at the forefront when

Michael Mandiberg (CC BY-SA 3.0).

renewable and zero-emissions energy sources become more widely used globally. Its goal is to grow economically by providing all of the technology for this type of energy.

Furthermore, China has an interest in reducing its use of fossil fuels like coal because they are causing thick smog in many Chinese cities. This smog makes people sick, reduces tourism, and makes the public frustrated with the government. China is already experimenting with some cap-and-trade schemes in cities suffering from smog problems. By March 2017, it had shut down all coal plants in Beijing. China is beginning a policy transformation that will improve the prospects of the global fight against climate change.

BriYYZ (CC BY-SA 2.0).

Thick smog blanketing the city of Shanghai.

Colombia

Shifting distributions of malaria cases in Colombia that are closely tied to temperature show how climate change could affect human health.

Colombia	
Population	47.2 million
GDP per Capita	US$14,100
Life Expectancy	76 years
CO_2 Emissions per Capita	1.7 metric tons

Colombia was part of a territory under Spanish colonial rule until it gained its independence in 1819. Throughout the nineteenth century, this territory broke up into multiple smaller countries until the Republic of Colombia as it is today emerged in 1903. Colombia is now the fourth largest country in South America by land area and the third largest economy on the continent. Over the past fifteen years, economic growth has improved living standards for many, but a large portion of its population lives in poverty. The country has highly varied geography, with densely populated mountainous regions in the northwest, tropical rainforests that are home to many different plant and animal species in the southeast, and low-lying coasts on both the Pacific and Atlantic Oceans. Its economy is based on exporting goods such as petroleum, coal, emeralds, coffee, bananas, and clothing. In fact, it is the fourth largest coal exporter in the world. Despite a history often marked by political conflict and violence, Colombia's economy is quickly growing, and it is becoming a more important player on the global stage. Studying how climate change affects Colombia can demonstrate how climate change affects human health.

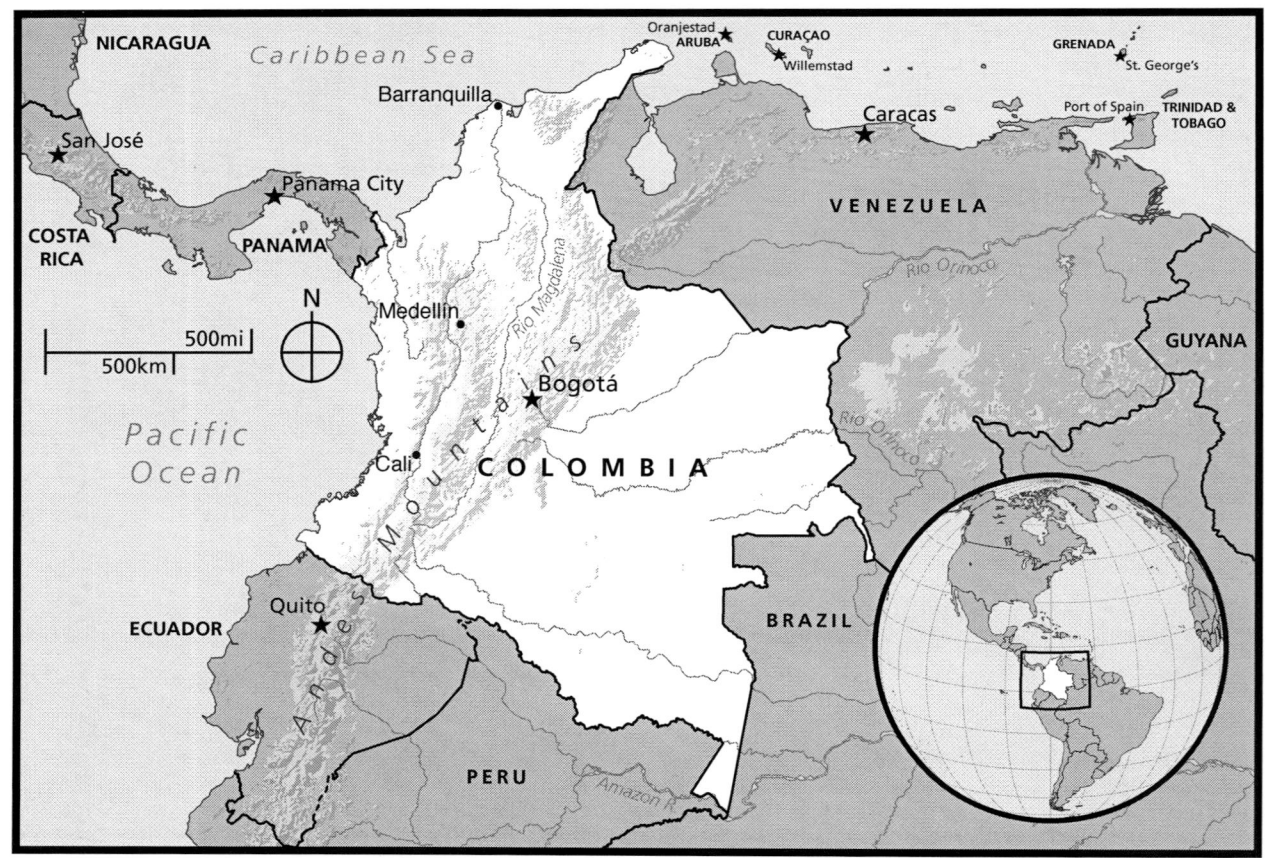

Colombian coffee is known as some of the best in the world. Certain types of Colombian coffee beans need to be hand-picked and hand-sorted, which makes it possible for small family farms to compete in the international market. Coffee plants are very sensitive to climate conditions, and climate change may be increasing the prevalence of disease among the plants.

How is Colombia experiencing climate change?

Because of the diversity of landscapes present in the country, Colombia is experiencing a wide range of climate change's effects. Between 1983 and 2013, increasing temperatures shrunk the glaciers (large masses of slowly moving ice) in the Colombian Andes Mountains by almost 60 percent. Today, these glaciers are melting even more quickly. From 2002 to 2007, Colombian glaciers shrunk, on average, by three square kilometers each year. At this rate, experts project that these glaciers will completely disappear before 2020. This affects the amount of water available for drinking and farming as well as for electricity generation (around 70 percent of the electricity produced in Colombia is from hydropower). In addition, like in many other coastal countries, flooding from sea level rise could displace many people and cause huge losses in the croplands that are vital to Colombia's economy.

Colombia may also experience some of the health effects of climate change. About 21 percent of Colombia's population is at risk of malaria infection, a disease that affects 300 to 600 million people around the world each year. Malaria can be deadly if not treated quickly, and in many parts of the world, malaria medicines are no longer effective to treat it. A certain type of mosquito spreads the parasite that causes the disease, and where these mosquitoes are able to live is highly dependent on climate conditions. While the environmental factors that determine where mosquitoes live are highly complex, both the malaria parasite and the mosquitoes that carry it generally thrive in warm temperatures.

For decades, the disease has been present only in the country's lower elevation regions, and its overall prevalence has even decreased throughout the country. Nevertheless, as temperatures have risen over the past few decades, more cases of the disease have emerged at higher elevations. With warming temperatures, the mosquitoes that spread malaria may be able to live in the higher, traditionally cooler, areas of the country—bringing malaria to those regions where it had not been common before. This means that climate change may put the dense populations of Colombia's mountainous regions at greater risk of malaria.

A rural road in Colombia. Colombia has a highly varied landscape including the Andes mountains, lowland coasts, and tropical rainforests.

James Gathany, Centers for Disease Control and Prevention's Public Health Image Library. ID #7861.

Certain types of mosquitoes, called Anopheles mosquitoes, can carry the malaria parasite and spread the disease to humans.

> *Our latest research suggests that with progressive global warming, malaria will creep up the mountains and spread to new, high-altitude areas.*
>
> —Menno Bouma, London School of Hygiene & Tropical Medicine, 2014

How is Colombia responding to climate change?

Before the mid-2000s, Colombia focused primarily on climate change mitigation projects that were also seen as economic opportunities. For example, it participated in programs that allowed wealthier countries to assist with projects, like new fuel-efficient bus systems, to reduce greenhouse gas emissions in Colombia. These projects helped Colombia pursue sustainable development while also helping wealthier countries reach their emissions reduction goals to prevent continued global climate change.

Colombia's response to climate change has recently shifted. In 2010 and 2011, the country experienced devastating rainfall, flooding, and cold from a cyclical climate condition called La Niña, which shifts ocean and air temperatures in the southern Pacific into a roughly ten month cold phase. The intense flooding affected four million Colombians and caused more than US$7 billion in damages related to loss of livestock, homes, and infrastructure. It is not clear how La Niña (and its warmer counterpart, El Niño) is linked to global warming, but the extreme weather made people in Colombia more concerned about the effects of a changing climate. After this environmental disaster, Colombia shifted its focus to adaptation. This is similar to other countries in South America, like Uruguay, where recent climate-related disasters led government officials and the public to pay more attention to climate change.

> *Colombia is not a country with high polluting emissions, but we want to assume our responsibility with the planet and its future.*
>
> —Colombian President Juan Manuel Santos, 2010

Colombia has now made climate change adaptation a national priority. Policies to combat the effects of climate change are integrated into many country-wide plans and projects. For example, Colombia created an Adaptation Fund to help farmers recover from the damaging effects of the 2010-2011 La Niña, and in 2017, began to implement its National Adaptation Plan for the UNFCCC. In addition, as part of the Colombian National Pilot Study of Adaptation to Climate Change, the country has increased its malaria monitoring and prevention efforts. Colombia has recently developed new maps, mathematical models, and early warning systems to help plan for changes in patterns of malaria exposure. Collaborations between government ministries that focus on climate and those that focus on malaria are increasing. Because of these efforts, Colombia has become internationally recognized as a leader in South America on climate change issues.

Freiburg, Germany

Careful city planning and changes in lifestyle in the city of Freiburg show what is possible when a highly developed urban community prioritizes reducing its emissions.

Freiburg is a wealthy city in the southwest of Germany that is home to many university professors. Nearly 12,000 of its 225,000 citizens work in environmental management or environmental science, including renewable energy research. With the average income in Freiburg almost 30 percent higher than the rest of Germany, the city generally has a high standard of living and has become internationally known for its environmentally friendly practices. For example, per capita CO_2 emissions from transportation in Freiburg are only 29 percent of the U.S. average.

Because Freiburg was heavily bombed during World War II, it had to rebuild much

Germany	
Population	80.7 million
GDP per Capita	US$48,200
Life Expectancy	81 years
CO_2 Emissions per Capita	8.9 metric tons

of its infrastructure after the war ended. This means that many of the buildings and roads in Freiburg today were constructed relatively recently, some with environmental concerns in mind. Freiburg serves as an example of how carefully-thought-out city planning can enable changes in lifestyle and make possible a highly developed urban community without harmful greenhouse gas emissions.

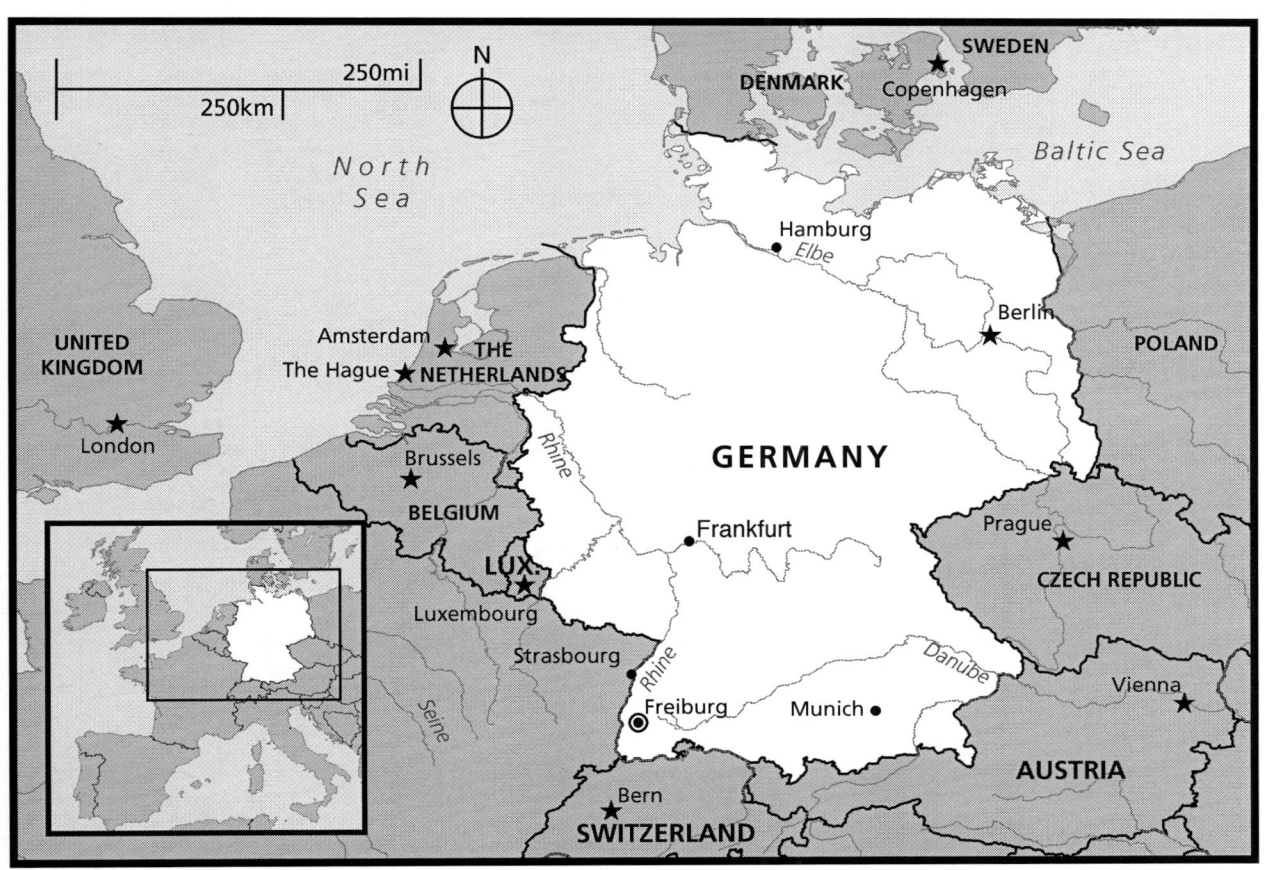

The "Solar Settlement" is a residential area in the Vauban quarter of Freiburg that uses solar panels to produce more energy than it uses. All the homes in this community are carbon neutral.

How is Freiburg experiencing climate change?

With over 1,800 hours of sunshine each year, Freiburg is one of the hottest, sunniest locations in Germany. Climate change is expected to increase the risk of heat-related health problems like heat stroke in southern Germany. There also may be more forest fires. In addition, fresh water will become more scarce in the region in the coming century, especially during the summer. On the other hand, Freiburg is within a major wine producing region of Germany, and global warming is expected to help the grape harvest. Overall, these environmental impacts are much less severe and immediate than in many other parts of the world. This makes adaptation less of a priority for the people of Freiburg, allowing them to focus more on mitigation strategies.

How is Freiburg responding to climate change?

Freiburg has worked to minimize its impact on the environment for decades. As early as 1996, city officials declared that Freiburg would strive to reduce its CO_2 emissions by 25 percent by 2010. Although it did not quite meet this goal, the city is now striving for a 40 percent reduction of emissions by 2030. Even more ambitious, it is hoping to become carbon neutral (contributing no additional CO_2 into the atmosphere) by 2050.

> ❝*Freiburg should not, nor does it want to, rest on its laurels, content with being a charming, engaging 'feel good city'.... Today, the city is also seen as a model combination of 'soft' ecology and 'hard' economy. Environment policy, solar engineering, sustainability, and climate protection concepts have become the mainstays of economic, political, and urban development.*❞
> —City of Freiburg Green City Office, 2011

Riding bicycles is a popular form of transportation in Freiburg.

Cora Went. Used with permission.

Easy access to public transportation in Freiburg makes it possible to have fewer cars in the city.

Freiburg is an international hub for research on renewable energy sources, especially solar. Solar panels are on many buildings—businesses, universities, private homes, churches, city hall, and even the soccer stadium. Some of the city's electricity is also produced by processing trash.

Some sections of Freiburg have gone even further in their efforts to reduce greenhouse gas emissions. The Vauban Quarter is a region of Freiburg built in 1998 that is close to the city center. There are no parking spots on the streets, and many people in Vauban do not have cars. (The city has an extensive public transportation system, and bikes are very popular.) The few people who do have cars must purchase expensive parking spots, which are very limited. Houses are specially designed to minimize energy needs, and cooperative living is common. This builds community among the residents, many of whom take pride in their environmentally friendly lifestyle.

While Freiburg has taken some dramatic steps toward sustainability, it is just one of many cities across the globe that is trying to reimagine how urban life and the natural world can more harmoniously coexist. This is especially important because by the year 2050, more than two-thirds of all people worldwide will live in cities.

Freiburg is a member of ICLEI-Local Governments for Sustainability, an international association of cities and local governments dedicated to sustainable development. While the ICLEI (International Council for Local Environmental Initiatives) is currently based in Germany, its member cities and towns are spread across the world in over eighty countries. This type of international organization helps create partnerships among cities so they can learn how other places are working toward reducing CO_2 emissions and building greater resilience against the effects of climate change.

New Orleans, Louisiana, United States

The destruction Hurricane Katrina caused in the city of New Orleans shows the dangers that extreme storms and sea level rise—both associated with climate change—pose to U.S. coastal cities. Hurricane Katrina also illustrates how poverty and long, deep histories of racism can amplify this danger and increase vulnerability to climate change.

United States	
Population	324 million
GDP per Capita	US$57,300
Life Expectancy	80 years
CO_2 Emissions per Capita	16.2 metric tons

New Orleans is a major port and Louisiana's largest city, with a population of over 380,000. The majority of the population is black. Located on a delta of the Mississippi River, New Orleans is sandwiched between Lake Pontchartrain to the north and the Gulf of Mexico to the south and east. Much of New Orleans lies several feet below sea level and depends on levees (natural or artificial ridges

made of soil and sand that line a body of water to prevent water from overflowing) and flood-walls to keep the city dry.

For years, New Orleans has struggled with high levels of poverty and racial segregation—as of 2015, 27 percent of the population was living in poverty, nearly twice the national average of 15 percent. Further, it is the city's black residents who are disproportionately af-

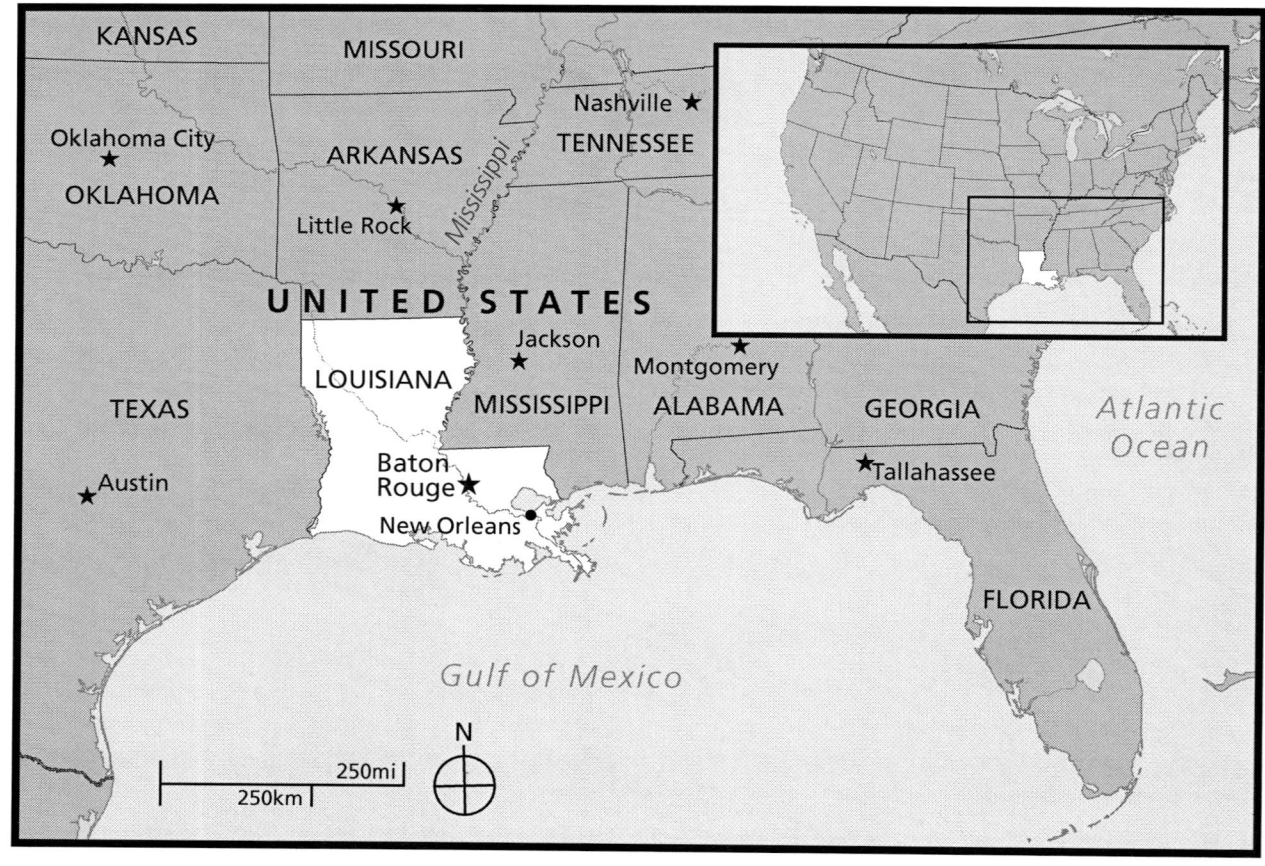

Infrogmation (CC BY 2.5).

A woman walks her dog along the levee next to the floodwall of the 17th Street Canal, which separates the city of New Orleans from the neighboring town Metairie. This photograph was taken a few months after Hurricane Katrina. Repair work to the floodwall on the New Orleans side of the canal is visible in the background on the right.

est European settlements were built on the higher elevation shores of the Mississippi River. But as the city expanded, people constructed homes at lower elevations. Although the city has a long history of confronting storms and floods, scientists predict that New Orleans will be increasingly threatened by changes in the global climate. Rising sea levels pose a growing challenge for the low-lying city, and scientists project that hurricanes coming off the warming waters of the Gulf of Mexico will be more frequent and severe.

In addition to its history of flooding, like many cities in the United States, New Orleans also has a history of racial segregation and structures that systematically disadvantage people of color and poor people. For example, certain policies, such as those related to housing, have segregated the city's residents by

fected by poverty. The city has become deeply segregated based on income with certain neighborhoods housing poorer residents and other neighborhoods housing the wealthy. Looking more closely at how New Orleans is experiencing and responding to climate change shows the dangers that U.S. coastal cities face—and how this danger is amplified for residents who are poor or people of color.

How is New Orleans experiencing climate change?

Experts have identified New Orleans as one of the cities in the world most vulnerable to rising sea levels. (Other U.S. cities that are especially vulnerable include New York, Miami, Los Angeles, and Seattle.) Since Europeans first arrived in the area more than three hundred years ago, New Orleans' coastal location and low elevation have made the city vulnerable to storms and floods. The earli-

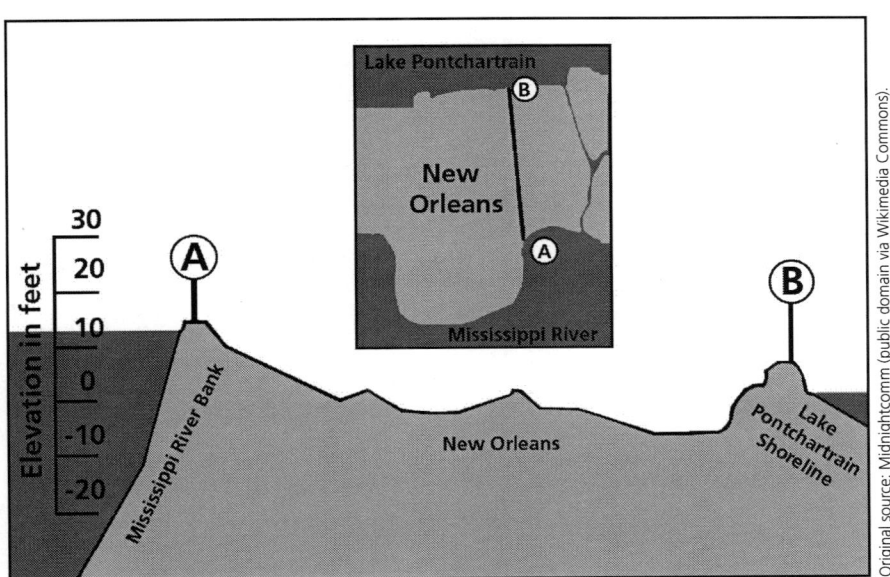

Original source: Midnightcomm (public domain via Wikimedia Commons). Modified by the Choices Program.

This graphic shows a cross section of New Orleans. Much of the city, which lies between two major bodies of water, is below sea level. While the vertical scale of this image is exaggerated compared to the horizontal scale, it shows how vulnerable the city is to flooding if its levees and floodwalls fail.

New Orleans residents wait to be rescued from the roof of their home as floodwaters from Hurricane Katrina surround them. Nearby houses are almost completely underwater.

race, often forcing them to live in areas vulnerable to flooding.

In August 2005, Hurricane Katrina devastated New Orleans. Levees and floodwalls were breached, and water surged into the city. About 80 percent of New Orleans flooded, with some places under more than twelve feet of water. The structural failure of the levee system, coupled with inadequate warnings and insufficient rescue operations, had disastrous effects for residents.

The hurricane claimed over 1,800 lives, mostly in the New Orleans area, and forced nearly 1.5 million people to evacuate. Within one month of the hurricane, people from New Orleans were seeking refuge in every state of the country. While Katrina caused intense flooding in most New Orleans neighborhoods regardless of race or wealth, areas stricken with poverty were least able to cope with the effects of the hurricane. In poorer, majority-black areas, people often did not have access to cars or other forms of transportation as the

hurricane approached, leaving many stranded and unable to escape. Twenty thousand people went to the Superdome, a stadium, for refuge, but the shelter was prepared for far fewer people.

> 66 *Each day, it was a battle to find enough food and water and get it to the Superdome. It was a struggle meal to meal because, as one was served, it was clear to everyone that there was not enough food or water for the next meal."*
> —Marty Bahamonde, a FEMA field officer who responded to the disaster, reflecting on the storm before Congress, 2005

Of those who died as a consequence of the storm, 67 percent were black, reflecting the breakdown of the population of New Orleans at the time. But many of the neighborhoods that were hit the hardest and where the most people died were predominantly black as black neighborhoods had been historically relegated to the low-lying areas of the city.

Although scientists generally do not attribute specific extreme weather events to climate change (and therefore do not say that climate change caused Hurricane Katrina), climate change will increase the frequency and intensity of storms like Katrina in the coming century.

The intensity of the storm, combined with the perceived failure of the government to respond adequately, led many to suggest that New Orleans and the United States were ill-equipped to deal with the effects of climate change.

> *What amazed many worldwide was that these extensive failures, often attributed to conditions in developing countries, occurred in the most powerful and wealthiest country in the world."*
> —Community and Regional Resilience Initiative Research Report, 2008

Members of the Congressional Black Caucus, the National Urban League, the National Council of Negro Women, the NAACP, and the Black Leadership Forum gathered after the storm to discuss the U.S. government response to Hurricane Katrina. They agreed that the government responded slowly and inefficiently, and they suggested that race played a major role in this response.

> *I feel that, if it was in another area, with another economic state and racial makeup, that President Bush would have run out of Crawford a lot quicker and FEMA would have found its way in a lot sooner."*
> —Reverend Al Sharpton, September 2005

To support the claim that the government's response has been dictated by racism, people have cited specific examples. For instance, the Federal Emergency Management Agency (FEMA) gave trailers to about 63 percent of

the residents of St. Bernard Parish—a majority white area. At the same time, FEMA only gave trailers to 13 percent of residents in the Lower Ninth Ward—a mostly black area also destroyed by the storm.

Many people across the country began to view the destruction from Hurricane Katrina as closely tied to racism in the United States.

> *We have an amazing tolerance for black pain."*
> —Reverend Jesse Jackson, in an interview, September 2005

How is New Orleans responding to climate change?

The severity of Hurricane Katrina and the destructive toll it took on the city of New Orleans sparked a national discussion about how to prepare for the city's future in the face of a changing climate. These conversations have largely focused on ways New Orleans can adapt to climate change, as opposed to how it can increase mitigation.

Local and national attention has focused on rebuilding the city to be better prepared for rising sea levels, stronger storms, and changes in precipitation patterns. Efforts include elevating existing buildings by several feet, constructing new escape routes through roofs, and improving emergency response plans. In the years following Katrina, Congress approved funding for multibillion dollar engineering and construction projects to build floodwalls, water pumps, and a chain of levees over one hundred miles long to protect New Orleans from future storms.

While many people have returned to the city, New Orleans' population as of 2014 was only about 85 percent of what it was before Katrina. Many areas of the city remain in ruins as construction and repair projects have been delayed.

Since the disaster, some activists and policy makers have worked to get the public and the government to acknowledge the role that racism and racist policies have played in increasing the vulnerability to climate change for populations of color. They have also called for policies that combat poverty and racial segregation in U.S. cities, including New Orleans.

The conservation and restoration of coastal wetlands are also important for increasing New Orleans' climate change resilience. For centuries, coastal wetlands (areas of swamps and marshes) provided a natural barrier to storms and flooding for residents of New Orleans. Today, the rate of the loss of wetlands in New Orleans is among the highest in world—on average an area of wetlands the size of one football field is lost every hour. Oil and gas industries have destroyed wetlands by dredging canals (deepening canals by scooping out mud and sand from the bottom) and building thousands of miles of pipelines in coastal Louisiana. Levees and dams on the Mississippi River have slowed the flow of sediment from the river that restores wetlands.

Many nongovernmental groups have pressed the state and federal government to do more to halt the destruction of these crucial wetlands, not only for the sake of wildlife, but for the sake of residents in New Orleans and the region. Some strategies would be to plant more cypress trees and marsh grasses along the coast as well as to divert water from the Mississippi River to replenish the wetlands with sediment. However, because these diversions would impact local people and businesses, determining where they should occur is challenging. Without significant action, much of coastal Louisiana could be underwater by the end of the century. This makes clear just how important local adaptation efforts are in the face of international leaders' failure to take strong action against climate change.

> **"Climate change is a threat that affects us all, and it is a real and present danger to our coastal communities. Here in Louisiana, we face a triple threat: subsidence [sinking land], coastal erosion, and sea level rise. If unchecked, New Orleans, like many coastal cities, will cease to exist."**
> —Mitch Landrieu, mayor of New Orleans, June 1, 2017

Haiti

Comparing the impact of Hurricane Jeanne on the Dominican Republic with the devastation it caused in Haiti illustrates how poverty increases vulnerability to climate change and reduces the ability to adapt.

	Haiti	Dominican Republic
Population	10.5 million	10.6 million
GDP per Capita	US$1,800	US$15,900
Life Expectancy	64 years	78 years
CO_2 Emissions per Capita	0.3 metric tons	2.1 metric tons

Haiti is one of two separate countries that occupy the island of Hispaniola in the Caribbean Sea. Haiti was a French sugar plantation colony populated largely by enslaved people that won its independence from France in 1804. The second country, the Dominican Republic, was a Spanish colony for three hundred years and became independent from Spain in 1820. Although these countries share the same island and have populations of around ten million people each, there are major differences between the two.

One of the most important differences is that Haiti is a much poorer country than the Dominican Republic. Nearly 60 percent of Haiti's population lives on less than $2.50 a day—a far higher number than in the Dominican Republic. A number of factors led to these differences in levels of poverty. International powers, such as Spain, France, and the United States have interfered in both countries' affairs throughout time and in different ways. For example, following Haitian independence, France ordered Haiti to pay it large sums of

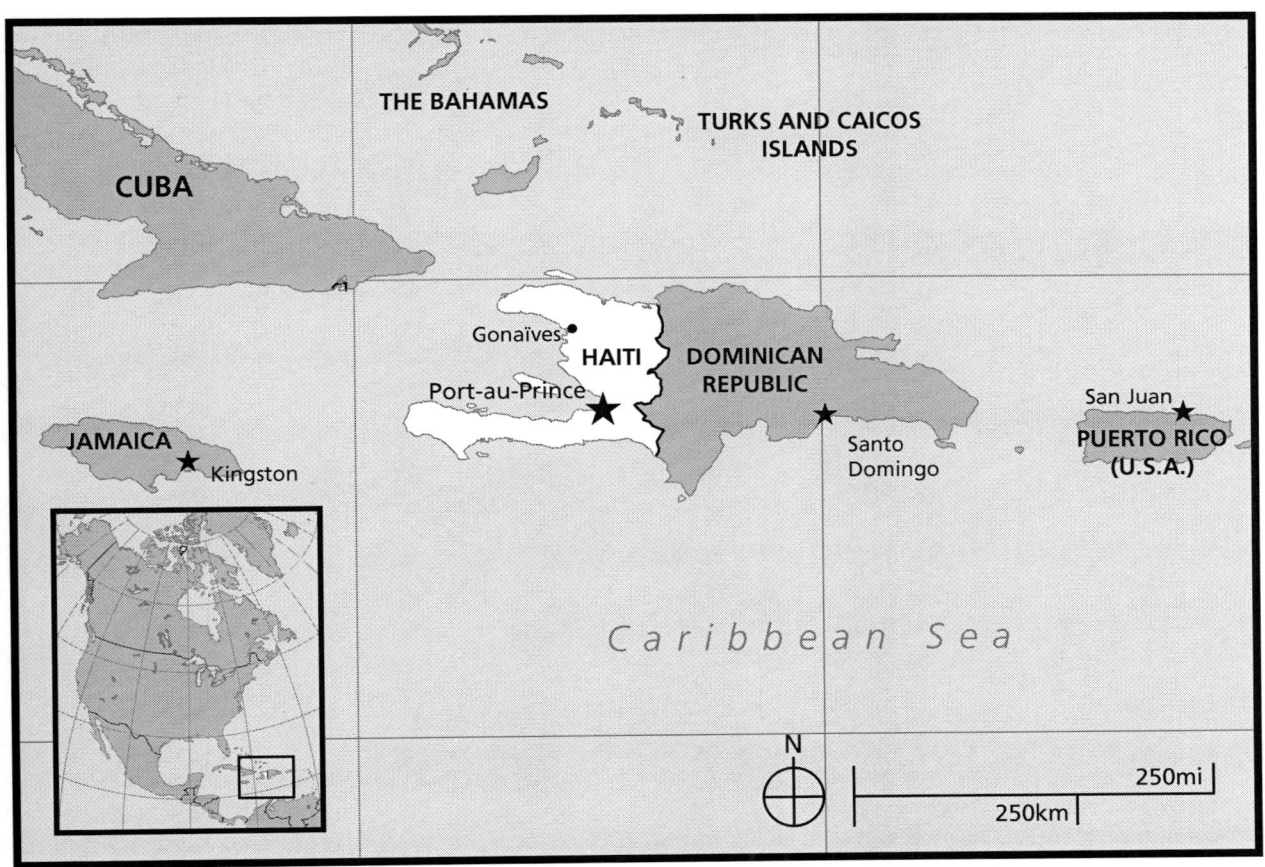

money or else it and other countries would not diplomatically recognize Haiti. Some scholars suggest that race and racism—Haiti is about 95 percent black—has also led international powers to discriminate against Haitians. These factors and many others set the stage for economic inequality between the two countries.

Comparing Haiti and the Dominican Republic helps show how poverty increases vulnerability to the effects of climate change. In addition, the case of Haiti illustrates how poverty can limit the capacity to adapt to climate change.

A woman stands in front of her home in Gonaïves, Haiti three days after Hurricane Jeanne. Many homes in Haiti are poorly built, crowded together in unsafe locations, and lack basic infrastructure like electricity and plumbing. Haiti's government does not have enough capacity to respond to disasters and depends heavily on international assistance.

How is Haiti experiencing climate change?

Sharp differences in the effects of extreme weather on Haiti and the Dominican Republic show how poverty can contribute to vulnerability of the population. The location of the island of Hispaniola in the Caribbean Sea places it in the path of hurricanes and tropical storms. Climate change contributes to more intense storms that come with strong winds and high levels of rainfall. When a powerful hurricane named Jeanne tore over the island in 2004, it killed more than three thousand people in Haiti but fewer than twenty-five people in the Dominican Republic. Flooding caused many of the deaths in Haiti. During the storm, torrential rains and walls of mud poured down Haiti's steep hillsides and collected in rivers that gushed into the city of Gonaïves where many of the deaths occurred. The storms destroyed crops, contaminated water supplies, and left more than 250,000 people in Haiti homeless.

A Haitian child carrying a food aid package from the U.S. military, which provided disaster relief after an earthquake hit Haiti in 2010.

A farmer growing cabbage in Haiti, where around 40 percent of the population works in agriculture.

Kendra Helmer, USAID.

>**❝***It was raining when we went to sleep. We were woken up by water in our beds, and in no time it was like an ocean invading us…. I heard my father calling for help. He couldn't move because he was handicapped. When I managed to get to his room, he was already taken by the water.***❞**
>
> —Nostra Gosette, a sixteen-year-old survivor of Hurricane Jeanne reflecting in 2005 on her experience

There is a direct link between poverty and the amount of damage that occurred in Haiti because of Hurricane Jeanne. It is because of this link that people in Haiti were more vulnerable than those in the Dominican Republic. Haiti's steep and hilly terrain was once covered with trees and rainforests that absorbed the rain and reduced the threat of mudslides caused by tropical downpours. Today, parts of Haiti are largely deforested because trees are used as the leading source of energy. Defor-

estation has put hillsides at a greater risk of soil erosion and the kinds of flash flooding that led to the deaths of so many people during Hurricane Jeanne. In contrast, because of the availability of electricity and other fuel sources in the Dominican Republic, the people there do not rely on harvesting trees for energy and do not face problems of such severity from deforestation.

>**❝***The crucial thing, because we're a country facing both an energy security crisis and a food security crisis, is how can we reconcile energy security and food security?***❞**
>
> —Gael Pressoir, Haitian nonprofit founder and business owner, November 2009

Constant erosion of fertile soil in Haiti makes planting trees to reforest the land more difficult. In addition, the fact that most farmers do not own the land they farm means that they often have little incentive to build embankments or to plant and protect trees to limit erosion and flooding. As erosion worsens, farmers face more challenges because the best soil has been washed away.

Farmers will also have to contend with other effects of climate change. Rising sea levels will lead to increased storm surges of saltwater in low-lying agricultural areas. Saltwater can contaminate fresh water sources and damage soil, making growing crops more difficult. Furthermore, higher temperatures and greater variation in rainfall patterns in Haiti have already led to more droughts during the dry season and more intense rain during the rainy season. These changes present challenges to farmers who lack access to information about new rainfall patterns and struggle to find the means necessary to adapt to changing growing seasons.

Looking to the future, reduced agricultural production due to climate change is one of the most serious issues facing Haiti. The country's economy depends on farming, and about 40 percent of Haitians work in agriculture. Damage to Haiti's farms and crops results in food scarcity and increases in the cost of basic goods for Haitians.

How is Haiti responding to climate change?

Haiti has developed a NAPA (National Adaptation Programme of Action) to address its vulnerability to climate change. But, as the poorest country in the Western hemisphere, it needs outside funding to implement the program.

The NAPA prioritizes strengthening the Haitian government's ability to anticipate and respond to the effects of climate change. For example, improving disaster preparedness is crucial, as is providing access to clean water sources and electricity. Other goals include reducing vulnerability to flooding through forest restoration, educating farmers about climate change and improving agricultural practices, and increasing living standards in rural areas. Ultimately, the most ambitious goal is to relieve poverty in order to reduce vulnerability to climate change.

While this process of planning is important, actually implementing a NAPA is particularly challenging in Haiti. Haiti's government struggles to provide even the most basic services to its citizens and depends on the UN, other countries, and many international NGOs for help with climate change adaptation. This creates confusion as different organizations sometimes fail to coordinate their activities or prefer to fund their own adaptation projects rather than those proposed by the Haitian government. Uncertain political conditions in Haiti have at times made international donors reluctant to contribute funds. Other issues also complicate the response. For example, many Haitian farmers, facing more immediate problems like daily hunger, delay learning about and adapting to climate change.

Poverty both makes Haiti more vulnerable to climate change and makes responding more challenging. This creates a cycle that is difficult to break.

Nigeria

Conditions in Nigeria bring to light how women are especially vulnerable to climate change and how they play an important role in resisting environmental degradation.

Nigeria is a multiethnic, multireligious country in West Africa. It is situated directly south of the Sahel (the transition area between the Sahara Desert and sub-Saharan Africa), and its southern border is located on the coast of the Gulf of Guinea.

Nigeria is one of the top oil producing countries in the world, processing more than two million barrels of oil per day. Competitions over profits from oil has caused a great deal of political and social conflict in Nigeria. The practices of oil companies like Shell and Chevron in the region have received criticism from local communities. Nigerian women have played an important role in protesting oil

Nigeria	
Population	186.1 million
GDP per Capita	US$5,900
Life Expectancy	53 years
CO$_2$ Emissions per Capita	0.6 metric tons

companies in the Niger Delta in some cases. In addition to shedding light on the tensions between powerful industries and their role in causing climate change, looking closely at Nigeria can help us understand the relationship between gender and responses to climate change.

How is Nigeria experiencing climate change?
The geographic variation in Nigeria means

Women attend a health education session in northern Nigeria.

that it is vulnerable to a wide range of effects of climate change. The North already experiences severe heat and water scarcity. Increasing temperatures will lead to extreme drought and heat-related illness, driving the people of this region into a state of desperation. In the South, high concentrations of people living along the coast are vulnerable to flooding as sea levels rise. Slums and other inadequate forms of housing are scattered along the most hazardous parts of coastal areas and wetlands. Here, poor people live in constant danger of flooding, particularly as their homes cannot withstand extreme weather. Poor people in Nigeria face some of the greatest vulnerability to climate change.

Throughout the world, poor women face this vulnerability to the extreme. In certain parts of Nigeria, this is also true. Many women in poor communities play a major role in the agriculture industry, farming in order to feed their families or to make a profit. These roles often expose women to greater impacts from climate change. For example, it is typical in rural areas for women to have the responsibility of collecting water for the household. As temperatures increase and water sources become more scarce, women will have to walk greater distances (often in extreme heat) to

access water. Women face these risks and others to a greater degree than men in poor communities throughout the world.

How is Nigeria responding to climate change?

The Nigerian government has some climate change-related policies. Local populations and nongovernmental organizations also play a vital role in responding to the threat of climate change. In the North, for instance, local communities are sharing insights on new agricultural practices as well as resources, such as seeds for growing crops, to help each other adapt to changes in climate. In the Niger River Delta, activists and local communities have also contributed greatly. Reflecting the relationship between gender and climate vulnerability, some of these groups have been influenced in important ways by women.

In the 1990s, communities in the Niger Delta (an area rich in oil reserves) began to express discontent with the practices of large oil companies and the Nigerian government. Local people disapproved of the way that oil profits were distributed. To further complicate the matter, oil corporations were also degrading the environment in the Delta. Most notably, they were illegally burning flares of natural gas during oil drilling. In Europe and North America, the natural gas that is released during oil extraction is collected and used in multiple ways, like for generating electricity or making chemicals. In Nigeria, however, oil companies burned the gas because this was cheaper. Burning natural gas in Nigeria has contributed more greenhouse gas emissions than the whole of sub-Saharan Africa combined. The flares also released toxic gases that threatened public health in the surrounding villages.

Tensions rose to new levels in 1995, when a group of activists from the Niger Delta were hanged by the military dictatorship after being charged with crimes that most people believed were fraudulent.

> **"Shell's day will surely come for there is no doubt in my mind that the ecological war that the Company has waged in the Delta will be called to question and the crimes of that war be duly punished."**
>
> —Ken Saro-Wiwa, Nigerian writer and environmental activist, on being sentenced to death by hanging, 1995

In addition to gas flaring, Shell and other companies have spilled an estimated 1.5 million tons of oil into the Niger Delta ecosystem over the past fifty years. This damages sensitive habitats for wildlife and contaminates local water sources, threatening the millions of people who rely on them for drinking, cooking, cleaning, bathing, and fishing. In January 2015, Shell agreed to pay more than $80 million to the Bodo community in the Niger Delta for the losses caused by two major oil spills in 2008 and 2009.

Later, during a nonviolent protest in 1999, police and government soldiers brutally attacked protesters and targeted the women who were involved. In general, women in the region also faced sexual violence at the hands of members of the Nigerian military. In response to the situation in the region, women's organizations banded together to fight the oil companies that were funding the government and its use of violence. During a ten month occupation of Shell's facilities, they demanded that flaring in Nigeria be completely phased out by 2007.

Throughout the early 2000s, groups in the Niger Delta—including some women's groups—and their international partners continued to demand changes in the practices of oil companies operating in the Delta. These women's groups were eventually able to persuade the Nigerian courts in 2006 to demand Shell end all flaring in the western part of the Niger Delta. In spite of this, the practice of gas flaring continues in Nigeria.

Nigerian protests against oil companies like Shell show that while in many cases women are particularly affected by climate change, they also have played an important role in resisting climate change and working for improvements.

Bangladesh

The determination of many poor local communities in Bangladesh to fight climate change shows the importance and effectiveness of locally-led adaptation.

Bangladesh is among the most densely populated countries in the world. One of the most prominent geographic characteristics of Bangladesh is the Ganges Delta, where the Ganges, Brahmaputra, and Meghna Rivers come together. This merging of the three rivers means that the area has richly fertile soil and expansive wetlands (areas of swamps and marshes).

The combination of this distinctive geography and the large Bangladeshi population has resulted in people living on *chars*, which are small river islands. While the economy of the country is based largely on manufac-

Bangladesh	
Population	156.2 million
GDP per Capita	US$3,900
Life Expectancy	73 years
CO_2 Emissions per Capita	0.5 metric tons

turing—with clothing and textiles as key exports—small-scale agriculture is the typical way of life on the *chars*. *Char* residents (as well as other rural Bangladeshi populations) live very interdependent and communal lives, often sharing resources and working together in the face of challenges. Because of this, Bangladesh provides many examples of locally initiated responses to climate change.

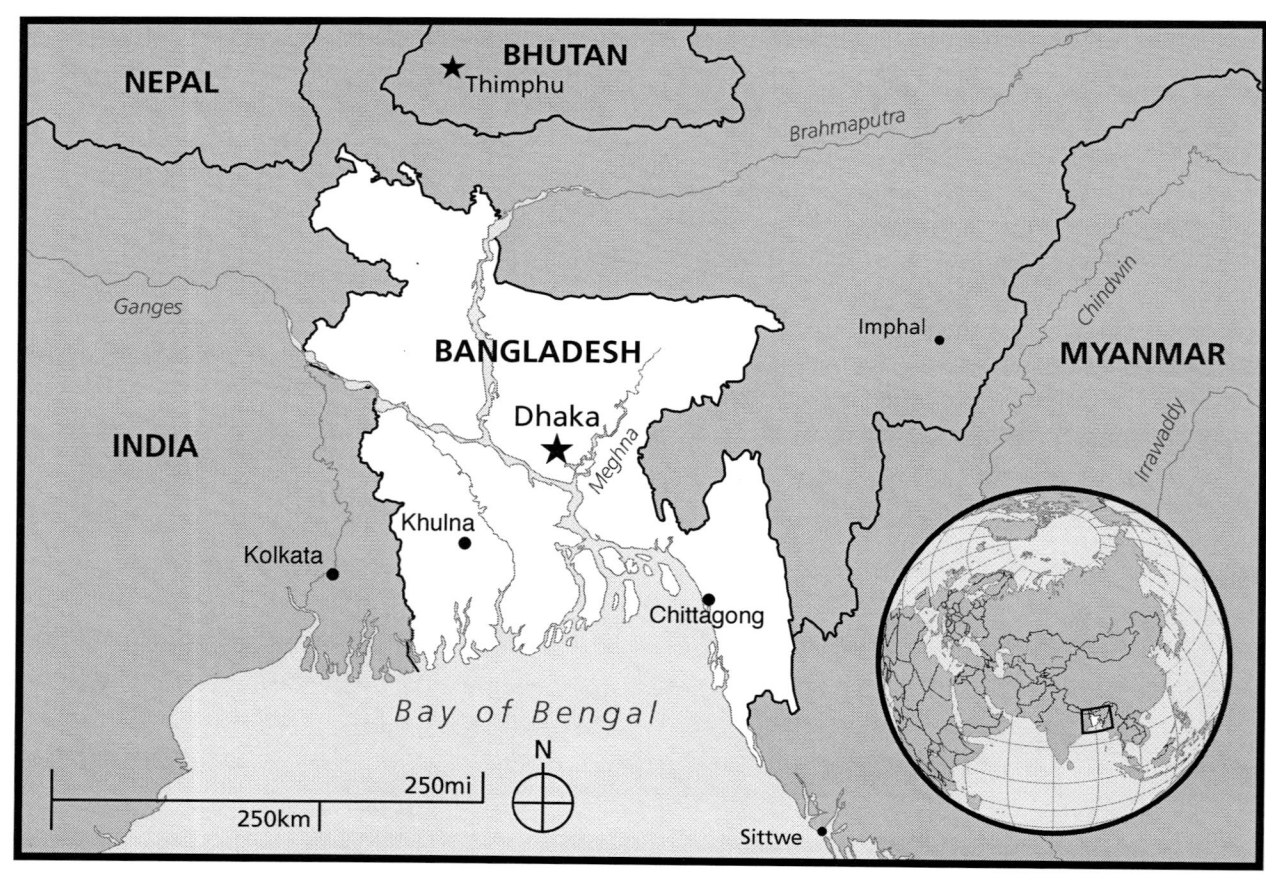

The low land elevation and prominence of rivers makes Bangladesh particularly prone to flooding.

How is Bangladesh experiencing climate change?

Because it is dominated by rivers and coastlines, Bangladesh is one of the world's most vulnerable countries to flooding. It is frequently hit by severe storms, and rises in sea level quickly envelop the already scarce and overpopulated land. Saltwater has already begun to seep into sources of drinking water and into farmers' soil, making growing certain crops such as rice much more challenging. As climate change causes rising sea levels and intensifying storms, Bangladesh will face more weather-related deaths, declines in public health, and too little land for a growing population.

Inhabitants of *chars* and wetlands are particularly vulnerable, and some of the islands are fully submerged for large parts of the year. This means that *char* residents have to migrate to other islands or the river shores until they can return to their homes. Their only other option is to move to cities, where urban slums are rapidly growing. Both in these quickly built slums and on the *chars*, there is no infrastructure for sanitation or clean drinking water. Waterborne diseases like cholera are common and spread quickly. Because the poorest people live in the most vulnerable areas, the effects of climate change hit them especially hard. These people are unlikely to be able to afford health care or to replace possessions lost in extreme weather.

Textile students at a school in Bangladesh.

Bangladeshi people building houses on stilts to protect their homes from flooding.

How is Bangladesh responding to climate change?

In Bangladesh, as in many countries in the global South, responses to climate change tend to rely on reducing poverty in order to decrease vulnerability. To do this, Bangladesh has focused primarily on education. Improving education helps increase Bangladesh's resilience to the effects of climate change in various ways. Better education allows for more career opportunities. This means that educated young people can pursue careers in sectors like manufacturing that are less sensitive to a changing climate than sectors like agriculture. Bangladesh has also promoted a link between education and migration. In attempts to reduce the pressure of a large population living on increasingly scarce land, Bangladesh has pursued what some call an "intentional brain drain." The goal is to educate Bangladeshi youth so that they can easily emigrate to other countries that need people with their qualifications.

In addition, the Bangladeshi government was one of the first to create a National Adaptation Programme of Action (NAPA). Yet the country's national government quickly realized that the NAPA would not be adequate in addressing the threat posed by climate change. As a result, Bangladesh created a more comprehensive national strategy and action plan for dealing with climate change. This plan included setting up two funding systems—one for wealthier countries to help fund climate change mitigation and adaptation efforts in Bangladesh and another for the Bangladeshi government to fund its own projects to counter climate change.

Many of the most effective adaptation efforts in Bangladesh have originated within poor communities. The dedication and knowledge of these communities have resulted in small scale adaptation efforts that have made Bangladesh a leader in the field. Ranging from floating farms that defy the restrictions of limited land to innovations in how homes are built to resist flooding, adaptation in Bangladesh has been an astounding example of community-led development.

In Kundetar Village, for example, a committee of local women has partnered with Gana Unnayan Kendra (a local NGO) and Oxfam to establish more effective disaster preparedness. These women have created a local network within their community for anticipating extreme weather events. They constantly listen to both Bangladeshi and foreign radio stations for any indication that bad weather could be approaching. If they do expect extreme weather, the women place important tools and resources in high places so they are accessible if flooding occurs.

An important benefit of locally-led adaptation is that it is defined by the very people who are experiencing the problems. This has not necessarily been the case with the national climate change response in Bangladesh. When drawing up its NAPA, the Bangladeshi government failed to adequately involve local populations in its deliberations. As a result, the priorities of the NAPA did not directly match the priorities of the Bangladeshi people, particularly the poor. Local efforts allow people to engage in adaptation in a way that aligns with their needs and interests. Local adaptation strategies also tend to be more easily shared because they are communal in nature and require the participation of all local inhabitants.

> **"It's difficult for the people of many countries like Bangladesh to face the double burden of poverty and impacts of climate change.... For the sake of sustainability of environment and development, we need to act now without delay, individually, locally, nationally, regionally, and globally."**
> —Prime Minister Sheikh Hasina of Bangladesh, 2013

Local adaptation efforts are not sufficient in the fight against global climate change. There are often funding shortages for the widespread use of adaptation techniques, and poor, local communities often lack the political power to demand climate change policy from the national government or international community. Furthermore, these populations often cannot implement many climate change mitigation strategies because they are responsible for so little of the world's greenhouse gas emissions.

Options in Brief

Option 1: Past Emitters Must Pay

Countries with long histories of greenhouse gas emissions must take responsibility for their harmful effects on the environment. Over a century of fossil-fuel driven industry in the global North has resulted in unprecedented spewing of greenhouse gases into the atmosphere. Climate change is a global problem, but its effects are most intensely felt by countries in the global South that have not contributed to the world's greenhouse gas emissions. Countries with histories of high emissions must accept strict limitations on fossil fuel use and provide funding to help those facing the effects of climate change. Countries in the global South should not have to face emissions restrictions that limit their development or bear the costs of adaptation. Promoting the economic development of poorer countries in this way will also help reduce their vulnerability to climate change's effects. Justice requires that we hold historic emitters accountable.

Option 2: Responsibility Must Be Shared by All

Climate change is a serious global problem, and it demands a global solution. We should encourage all countries to follow through on the pledge that they made in the Paris Agreement to reduce their emissions. Reducing fossil fuel use will shift international attention to sustainable development in both wealthy and poor countries. Sustainable development will allow countries around the world to meet the immediate economic and social needs of their citizens without compromising the future. We cannot let the short-term economic costs of establishing new energy infrastructure outweigh the long-term benefits of sustainable development in mitigating a global climate crisis.

Option 3: Economic Growth and Development Must Come First

Climate change is not our most immediate problem. Billions of people around the world face poverty and hunger every day. How can we justify an international focus on expensive climate change mitigation and adaptation strategies when this is the case? In order for the health and education of people in the global South to improve, international attention must turn toward economic development and poverty reduction. Most importantly, all countries should be able to grow their economies by increasing their industrial activity. Continuing to hope for global climate change mitigation is unrealistic—it is too expensive and people will not give up a higher quality of life for the sake of reducing emissions. Wealthy countries should instead encourage businesses and organizations to focus on scientific research to develop technologies that can help us deal with the effects of climate change. If countries each pursue their own economic growth, the market will generate solutions to the problem of global warming.

Option 1: Past Emitters Must Pay

Countries with long histories of greenhouse gas emissions must take responsibility for their harmful effects on the environment. Over a century of fossil-fuel driven industry in the global North has resulted in unprecedented spewing of greenhouse gases into the atmosphere. Throughout history, the United States has dumped more than three times the amount of carbon dioxide (CO_2) into the atmosphere than newly industrializing countries such as China and well over one hundred times the amount emitted by many poorer countries such as Bangladesh.

> **❚❚Climate justice endorses that polluters must pay. We must have a system that those who use SUVs, not the one[s] who use bicycles, pay."**
>
> —Kofi Annan, former Secretary General of the United Nations, 2014

Climate change is a global problem, but its effects are most intensely felt by countries in the global South that have not contributed to the world's greenhouse gas emissions. The effects of climate change are costly—financially and in human suffering. Preventing further damage requires large investments in adaptation projects.

Countries with histories of high emissions must accept strict limitations on fossil fuel use and provide funding to help those facing the effects of climate change. Countries in the global South should not have to face emissions restrictions that limit their development or bear the costs of adaptation. They should have the opportunity to industrialize and gain an equal footing in the global economy. Promoting the economic development of poorer countries in this way will also help reduce their vulnerability to climate change's effects. Justice requires that we hold historic emitters accountable. Any alternative would not be fair to the people in the global South who are suffering first and worst from climate change.

Option 1 is based on the following beliefs

• Climate change is one of the most serious challenges facing our planet. People who claim that climate change is not real threaten our ability to address the problem.

• Justice requires accountability for past wrongs.

• Those most responsible for causing climate change must bear the costs of solving it.

• Major reductions in greenhouse gas emissions by historic emitters will be enough to prevent climate change.

• International agreements about climate change should prioritize addressing the needs of the countries most vulnerable to the effects of climate change.

What policies should we pursue?

- Historic emitters must accept mandatory emissions restrictions and pursue sustainable development.

- Historic emitters must pay for adaptation efforts—including agricultural assistance, disaster preparedness, and health care improvements—in countries most vulnerable to climate change.

- The global South should be allowed to develop without emissions restrictions.

Arguments for

1. As a global community, we cannot ignore the past emissions that have brought us to the edge of catastrophic climate change. For an international system to be grounded in justice, it must hold countries accountable for historical emissions.

2. Countries with low emissions records are suffering the consequences of climate change and do not have the ability to mitigate or adapt. These countries are entitled to assistance from those who caused the problem.

3. Poor countries trying to reduce the small amounts of greenhouse gases they emit will not make a significant difference. For mitigation to be effective on a global scale, the countries that have emitted the most over time must take the lead.

4. Historically responsible countries have reaped the benefits of decades of industrial growth with little regulation; they can afford to pay for mitigation and adaptation (both at home and in other countries that need it most).

5. Poor countries need a chance to develop economically and must be either allowed to increase their greenhouse gas emissions or helped by wealthy countries to pursue sustainable development. Economic development will also reduce poorer countries' vulnerability to climate change.

6. International negotiations that result in mandatory rules and restrictions are the only way to make meaningful progress in preventing and dealing with climate change.

Arguments against

1. The United States will not agree to international treaties that include emissions cuts or that require it to contribute funding to poorer countries' adaptation programs. Without U.S. participation, any large-scale attempts at mitigation will not be effective in curbing emissions.

2. China has the largest total emissions of any country. Its use of fossil fuels will only increase in the future. Without restrictions on rapidly industrializing countries like China and India, an international system would not prevent future climate change.

3. Restricting the emissions of some countries but not others will give the latter an advantage in the global market. This is unfair to citizens of countries facing restrictions, which will almost certainly lose jobs to other countries that will take the lead in industry.

4. The citizens of historically high-emitting countries are not as directly or immediately vulnerable to climate change's effects. Leaders of these countries will not be able to justify sweeping emissions restrictions to their citizens. These countries are also the most powerful in the international community making it is unlikely that an effective agreement will be reached without their support.

Option 2: Responsibility Must be Shared by All

Climate change is a serious global problem, and it demands a global solution. We need to be practical about the types of emissions reductions the most powerful, highest emitting countries will realistically agree to in order to make any progress in preventing the dangerous effects of climate change.

China emits more greenhouse gases than any country in the world. Many other countries will greatly increase their greenhouse gas emissions as they continue to industrialize. Why should these countries be allowed to continue damaging the environment? We should encourage all countries—both rich and poor—to follow through on the pledge that they made in the Paris Agreement to reduce their emissions. This approach will be most effective in preventing future greenhouse gas emissions and will also make countries like the United States more likely to sign on. In addition, wealthier countries will not feel that they are at risk of falling behind in the international market. As a result, they will be more likely to contribute funding to help the countries most affected by climate change to adapt.

> **❚❚ Emissions are rising fastest in emerging economies and in the interest of their poorest citizens on the front line of climate change, they must play a bigger role than in the past."**
> *—Jan Kowalzig, Oxfam's climate expert, 2014*

At the same time, we cannot deny the importance of development to support economic growth and increase climate change resilience around the world. Reducing fossil fuel use will shift international attention to sustainable development in both wealthy and poor countries. Sustainable development will allow countries around the world to meet the immediate economic and social needs of their citizens without compromising the future. We cannot let the short-term economic costs of establishing new energy infrastructure outweigh the long-term benefits of sustainable development in mitigating a global climate crisis.

Option 2 is based on the following beliefs

• Climate change has already begun to have catastrophic effects on people around the world. Denying or ignoring it is irresponsible.

• Justice requires that we stop current wrongs and prevent future harm.

• This global problem requires that everyone takes responsibility, with those who are most able to pay giving financial support to others.

• Voluntary emissions restrictions for all countries is the only way to reach an agreement that will involve the entire global community and prevent climate change.

• All countries and organizations should have equal participation in negotiating agreements on global climate change.

What policies should we pursue?

• There should be across-the-board voluntary emissions restrictions proportional to countries' current emissions levels. This will encourage both the global North and the global South to reduce future greenhouse gas emissions.

• We should encourage all countries to focus on sustainable development and production.

• Wealthier countries should be encouraged to voluntarily provide funding for adaptation efforts in poorer countries.

Arguments for

1. U.S. or European cuts would be pointless if China, India, and other countries in the global South continue to increase emissions.

2. Reducing the focus of blame on the global North will make wealthier countries more likely to contribute funds for adaptation efforts.

3. Emissions restrictions for all countries provides the most realistic chance of reaching an international agreement and will be the quickest route to concrete global action. The longer we wait, the more expensive mitigation and adaptation will be.

4. Across-the-board emissions restrictions will encourage sustainable development, allowing countries to improve economically without endangering future climatic conditions.

5. To protect the future state of the environment, we must prevent future greenhouse gas emissions. Climate change poses a dire threat to life as we know it, and we must do all we can to prevent catastrophe.

6. International negotiations provide the only chance to prevent climate change and realize justice on a global scale.

Arguments against

1. Poor countries deserve the chance to industrialize the way that the United States and other wealthier countries did. Restricting their use of fossil fuels would prevent this.

2. The greenhouse gases in the atmosphere causing climate change are from the past. We must hold industrialized countries accountable for this.

3. Some countries already need to pursue expensive adaptation efforts in the face of their vulnerability to climate change. They should receive financial assistance from the wealthy countries that caused climate change. Funding should be mandatory. Voluntary contributions will not be enough.

4. Limiting the economic development of countries in the global South by regulating their emissions will slow poverty reduction. Continued poverty will keep these countries vulnerable to the effects of climate change.

Option 3: Economic Growth and Development Must Come First

Climate change is not our most immediate problem. Billions of people around the world face poverty and hunger every day. How can we justify an international focus on expensive climate change mitigation and adaptation strategies when this is the case? In order for the health and education of people in the global South to improve, international attention must turn toward economic development and poverty reduction. Most importantly, all countries should be able to grow their economies by increasing their industrial activity.

There should be no mandatory restrictions on fossil fuel use. All countries should have the right to industrialize cheaply.

> **❚❚It is prudent to do what we reasonably can to reduce carbon emissions. But...we don't believe in harming economic growth.... For many decades at least, [fossil fuels] will continue to fuel human progress as an affordable energy source for wealthy and developing countries alike."**
> —*Prime Minister Tony Abbott of Australia, 2014*

Individual countries, as well as local organizations, can choose to take on voluntary emissions restrictions and adaptation measures. Economic development will also allow poorer countries to reduce their vulnerability to climate change. This approach may also be faster than holding out for a comprehensive global agreement on climate change. Any attempts by the United Nations (UN) to barge in and tell people how to adapt to climate change will fail. The UN tramples on local governments and ignores citizens' input.

Continuing to hope for global climate change mitigation is unrealistic—it is too expensive and people will not give up a higher quality of life for the sake of reducing emissions. This makes reaching an international agreement on mitigation essentially impossible. Wealthy countries should instead encourage businesses and organizations to focus on scientific research to develop technologies that can help us deal with the effects of climate change. If countries each pursue their own economic growth, the market will generate solutions to the problem of global warming.

Option 3 is based on the following beliefs

• Climate change is taking place, but it is not a serious threat that requires drastic action.

• Justice is allowing everyone the opportunity to pursue prosperity.

• The international community is not responsible for funding and promoting mitigation and adaptation. These concerns should be dealt with by national governments of individual countries according to the interests of their populations.

• Businesses should have a key voice in considering climate change policies. In a thriving economy, technological solutions will be developed that can eliminate the harmful effects of climate change.

• Policies generated by local people will be more successful than those forced upon them by international leaders who are not engaged with local concerns.

What policies should we pursue?

• Local and national governments as well as businesses and other nongovernmental groups can choose to take on voluntary emissions reductions, but there should not be mandatory emissions restrictions.

• The international community should encourage countries to pursue economic growth so they have money available to research and develop new technologies—both renewable energy options and geoengineering techniques that could help deal with the effects of climate change.

• Monetary aid from wealthier countries should continue to focus on the economic development of poorer countries, not on climate change adaptation.

Arguments for

1. The most catastrophic effects of climate change are not as immediate or important as the economic needs of people. Economic development will give governments more money to fund public health initiatives, improve education, and reduce poverty. Expensive attempts at climate change mitigation would cripple countries' economies, making less funding available to deal with these pressing issues.

2. Climate change prevention is too expensive. Realistically, it will never happen.

3. Countries have a right to develop, and people have a right to pursue prosperity. This requires the use of fossil fuels. It is unfair to deny more than half the world the benefits of industrial development.

4. Restricting development increases vulnerability to climate change by preventing both poverty reduction and the creation of diverse job opportunities.

5. Technology has the potential to prevent the most dangerous effects of climate change. We should make sure we have the money to pursue innovative technologies to counteract these effects.

6. International climate change talks have failed to produce meaningful agreements that all governments can agree with, and international laws restrict the ability of countries to develop in ways that are suited to their unique local contexts. It is unrealistic to rely on the international community to effectively protect the interests of all peoples.

Arguments against

1. Pursuing sustainable development would allow countries to prioritize the economic needs of their peoples and address issues like health, education, and poverty while also limiting greenhouse gas emissions.

2. We cannot rely on hopes of developing new technology to "solve" climate change. People are already suffering as a result of global warming. We cannot wait any longer to take action. Furthermore, geoengineering is dangerous, cannot be tested, and does not address the source of the problem.

3. Pursuing efforts to reduce greenhouse gas emissions is essential. Unchecked industrialization could have disastrous environmental effects, potentially resulting in many parts of the world becoming uninhabitable. Long-term human costs outweigh short-term economic expenses. In addition, proactive mitigation strategies will save money in the long run by preventing future damages and lessening the need for expensive adaptation efforts.

4. Because climate change is a global problem, solving it requires a coordinated, international effort. Voluntary action is not enough because the people and groups that are motivated to act often are not the ones who have caused the problem.

Supplementary Resources

Books

Kolbert, Elizabeth. *Field Notes from a Catastrophe: Man, Nature, and Climate Change.* New York: Bloomsbury Publishing USA, 2006.

Meckling, Jonas. *Carbon Coalitions: Business, Climate Politics, and the Rise of Emissions Trading.* Cambridge, MA: The MIT Press, 2011.

Newell, Peter and Matthew Paterson. *Climate Capitalism: Global Warming and the Transformation of the Global Economy.* New York: Cambridge University Press, 2010.

O'Brien, Karen, Asunción Lera St. Clair, and Berit Kristoffersen, eds. *Climate Change, Ethics and Human Security.* New York: Cambridge University Press, 2010.

Roberts, J. Timmons and Bradley Parks. *A Climate of Injustice: Global Inequality, North-South Politics, and Climate Policy.* Cambridge, MA: The MIT Press, 2006.

Roberts, J. Timmons, David Ciplet, and Mizan Khan. *Power in a Warming World: The New Geopolitics of Climate Change.* Cambridge, MA: The MIT Press, Fall 2015.

Ruddiman, William F. *Plows, Plagues, and Petroleum: How Humans Took Control of Climate.* Princeton, NJ: Princeton University Press, 2010.

Online Resources

Energy Information Administration <http://www.eia.gov>. Federal agency that provides current and historical data on energy production and use as well as the environmental effects of energy consumption.

Global Governance Monitor: Climate Change <http://www.cfr.org/global-governance/global-governance-monitor/p18985#!/climate-change>. A comprehensive educational webpage produced by the Council on Foreign Relations about how the international community is responding to climate change.

Intergovernmental Panel on Climate Change <http://www.ipcc.ch>. The leading international body for scientific, technical, and socioeconomic climate change data and analysis.

Risky Business <http://riskybusiness.org>. A project and formal report focused on quantifying and publicizing the economic risks from the impacts of a changing climate.

U.S. Global Change Research Program <http://www.globalchange.gov>. Federal research program that produces the National Climate Assessment, supports policy decision-making, and provides tools for educating about climate change.

UN Climate Change Newsroom <http://newsroom.unfccc.int>. Website of the United Nations Framework Convention on Climate Change (UNFCCC) with the latest international news on climate change research and responses as well as information about UNFCCC meetings and process.

What Next Volume III: Climate Development and Equity <http://www.whatnext.org/Publications/Volume_3/Volume_3_main.html>. An online collection of articles by commentators, researchers, and activists around the world covering climate change through the lens of equity.